Cambridge Atmospheric and Space Science Series

Editors

Alexander J. Dessler
John T. Houghton
Michael J. Rycroft

Titles in print in this series

M. H. Rees
Physics and chemistry of the upper atmosphere

Roger Daley
Atmosphere data analysis

Ya. L. Al'pert
Space plasma, Volumes 1 and 2

J. R. Garratt
The atmospheric boundary layer

J. K. Hargreaves
The solar-terrestrial environment

Sergei Sazhin
Whistler-mode waves in a hot plasma

S. Peter Gary
Theory of space plasma microinstabilities

Martin Walt
Introduction to geomagnetically trapped radiation

Tamas I. Gombosi
Gas kinetic theory

Boris A. Kagan
Ocean–atmosphere interaction and climate modelling

Ian N. James
Introduction to circulating atmospheres

J. C. King and J. Turner
Antarctic meterorology and climatology

J. F. Lemaire and K. I. Gringauz
The Earth's plasmasphere

Daniel Hastings and Henry Garrett
Spacecraft–environment interactions

Thomas E. Cravens
Physics of solar system plasmas

John Green
Atmospheric dynamics

Atmospheric Dynamics

John Green

CAMBRIDGE
UNIVERSITY PRESS

PUBLISHED BY THE PRESS SYNDICATE OF THE UNIVERSITY OF CAMBRIDGE
The Pitt Building, Trumpington Street, Cambridge CB2 1RP, United Kingdom

CAMBRIDGE UNIVERSITY PRESS
The Edinburgh Building, Cambridge CB2 2RU, UK http://www.cup.cam.ac.uk
40 West 20th Street, New York, NY 1011-4211, USA http://www.cup.org
10 Stamford Road, Oakleigh, Melbourne 3166, Australia

First published 1999

Printed in the United States of America

Typeset in Times 10.25/13.5pt, in 3B2 [ᴋᴡ]

A catalogue record for this book is available from the British Library

Library of Congress Cataloguing-in-Publication Data

Green, John (John Sydney Adcock), 1934–
    Atmospheric dynamics / John Green.
        p. cm. – (Cambridge atmospheric and space science series)
    Includes index.
    ISBN 0 521 24975 9
    1. Dynamic meteorology. 2. Atmospheric physics. 3. Fluid
    mechanics. I. Title. II. Series.
    QC860.G66 1999
    551.5′ 15–dc21   97-43007 CIP

ISBN 0 521 24975 9 hardback

Contents

x   Contents

Introduction

This volume is the condensation of lectures given to students at Imperial College Department of Meteorology, while it existed, and to students of Environmental Science at the University of East Anglia when the department at Imperial College was closed down. Any good sense it might contain is almost certainly attributable to colleagues. Eric Eady opened my eyes to many fascinating phenomena, though when I taught students I began to realise that such communication is perhaps more of a two-way process than I imagined at the time. It is one thing to speculate on an interpretation, but another when this stimulates a response. Recalling a technical difficulty to a non-technical partner is perhaps a simple illustration of the suggested interaction. There is also a lot of Frank Ludlam here. He asked for explanations of mathematical theories that he could understand without having to go through the detailed mathematics. I remember an early encounter when he described his method of evaluating integrals. Simple, he said, 'you just move the variables, one by one, through the integral sign 'till you are left with an integral that you *can* do'. Aspects of this collaboration can be seen in his book on *Clouds and Storms*, which has some of me in it, but which for political reasons was forgone; Frank died before his wonderful book was published. The philosophy here is almost all that of Professor Peter Sheppard, who tormented, teased, threatened, sometimes disillusioned, many generations of students, for his criticisms were accurate and acid. If your theory could stand up to him, it would stand up anywhere. 'Where's your nose Green' I still hear him say, for he knew when innovation was essential too. Finally there were many students, and occasionally I recollect that it was a student who made some nice observation, not me. I distinctly remember one, and I remember who, when asked what we knew about ozone said 'ignites with a glowing taper'. A beautiful comment on the inadequacy of our textbooks to prepare us for the environment.

Collaboration with diverse colleagues and students encourages a degree of interpretation that is contrary to much mathematics. However, when you give up proof, in favour of suggestion, it allows you to see what you have gained in the proof. It is common experience that as a proof becomes more rigorous, the domain to which it can be applied becomes more restricted. I think there is a role for more general, less specific analysis. This reduces me to being dissatisfied with most of the things I have published, except where I have been able to say that they have led to some sort of tangible enlightenment. Thus you will find even quite trivial manipulations leading to speculation as to another interpretation.

There are essentially no references here. Whenever I tried to read the literature I would quickly go off on an interesting tangent, and never get to the end of the article to be able to incorporate it in my vocabulary. That is a bad thing for a student to do, but it might be more forgiveable than to read up a subject before studying it. One is readily convinced by the word of the current authority, and it always seemed to me to be much more fun to go off on some untrodden path than to compete with current researchers exploiting the latest theory trying to get the next result first. It is also important to have prior feeling about the topic you are reading about to compare with what the author is telling you. They may be telling you something you already know in incomprehensible language. An unbiassed opinion is almost as irrelevant as random motion.

There is not much in the way of illustrative material either. It is so easy to say 'here is an example of the phenomena being described' upon which the attentive student goes away with this thumbprint. What really happens is that there never is such a good example, all the real ones go off imperfectly, and there is no typical example. Thus I would rather the student had the notion that there were some stylised things that real phenomena might remind him of. As an example of deception, there are two speculations about wave propagation in the atmosphere. One is that influences are propagated downstream, the other that they are propagated along great circles. If we draw a polar projection, we see propagation along great circles, and in an equatorial projection, latitudinal trains of waves.

Meteorology has reached a complex position. We have access to a great bank of data and the computing power to analyse it. I came into meteorology through 'there is a wind on the heath brother, all good things', and am inspired by that feeling, and it seems to me that we are in danger of being unable to follow through from that feeling to modern data and analysis. I hope some people will be able to savour, more acutely, that connection.

Chapter 1

Description of atmospheric motion systems

And God created great whales, and every living thing that moveth, which the waters brought forth abundantly after their kind, and every winged fowl after his kind: and God saw that it was good.

1.1 Introduction

We usually notice phenomena; events isolated in space, but most of this book will rely on a wave formalism. This chapter explores some of the relations between the two forms of description. We are going to describe a great variety of motion systems, with broad classes, but with each member of each class different. Moreover, the definition of the class depends, to some extent, on the reason we are trying to classify the phenomenon. A cumulus cloud is fairly well defined, in the sense that two independent observers will (usually) agree that a specific cloud should be put in that broad category. There will be some debate as to whether this specimen is young or old, becoming congested, and such subtleties are important as indicators of the future development of the convection. The particular one we see now is an individual, and studied for a specific purpose; it might make a gust that will disturb my boat, cover up the sun, precipitate soon, carry aphids/momentum/water vapour, into the higher levels of the troposphere; a whole host of things that will affect the way I look at it.

Usually, we are concerned with relations between things; like the organisation that causes cumulonimbus to occur in particular regions of a frontal zone; frontal zones to appear in fairly well defined regions of a weather system; weather systems appear in certain zones of the circulation of larger scale. We become aware of the importance of the spaces between the more obvious parts of a system.

It is comforting to see individual motion-systems. A depression, with its region of closed isobars on the surface chart appears quite well defined, and we can therefore speak of its position now, and where it and its associated weather will be tomorrow. But on the 500 mb chart, which is more representative of the bulk of the atmosphere, the closed isobaric centre is usually absent, and is replaced by a trough in the streamlines which is less well defined and less spectacular. Moreover, the air currently near (including above) the surface low, is moving in response to forces that have been organised for a few days, during which time air has travelled several thousand kilometers from a wide variety of directions. In some sense, the scale of this system is, at least for some purposes, very much larger than that of its most obvious surface feature.

The beautiful crisp edge of a cumulus cloud defines the limit of penetration of potentially warm moist air from below. Clear air must be moved out of its way (by the action of the pressure field) so there must be descent of dry, negatively buoyant, air outside the cloud. This is an important aspect of the circulation, for it is this descent which leads to the general rise in temperature of the layer of air into which the cloud penetrates. Moreover it is one mechanism by which one cloud interacts with another to give organisation on the scale of the cloud field. This way we get the notion that, at least for some purposes, motion systems occupy all the available space. For them, spectral analysis, using for example sinusoidal functions (Fourier series) might be useful. These allow uniform representation all over space, uniquely separate scale from intensity, and help us to identify some relevant physical processes.

Sinusoidal functions are not so convenient for describing systems that are truly isolated in space. For example a single updraught of width a needs all wavenumbers up to several times a^{-1} together with their phase relations, for its description. Here we will be concerned mainly with the long-time evolution of the atmosphere, therefore with the interaction beween smaller and larger scales of motion. Thus individual motion systems acquire a rather transient nature and we tend to regard the wavelike description as more fundamental than the phenomenological even though sometimes less fun. The contrast between the 'wave' and 'particle' description of light, where some phenomena are more easily described in terms of one idealisation rather than the other, is a useful analogy.

1.2 Spectrum of motion

The vast majority of the kinetic energy of the atmosphere is contained in the zonal mean of the zonal component of the flow. This represents a sort of flywheel, in which energy is stored and may be available for other motion systems to make use of, but which itself does little. In broad terms the wind increases from being small near the ground to an average of some 20–30ms^{-1} in the same sense as the rotation of the earth, near the tropopause, especially in middle latitudes. See Figure 15.1 and the discussion of Section 15.2 for further details. Superimposed on this are wavelike systems, on which we now direct our attention.

Motion of wavelengths from a few mm, severely affected by viscous forces, to the 40 000 km of the size of the earth are of interest to us. Because of this great range of scales it is convenient to use a logarithmic scale for the wavelength, and adjust the amplitude scale so that it still represents energy. The log-scale has the additional advantage that we can equally well label the axis with spatial wavenumber, which is usually a more convenient mathematical parameter than wavelength, and with some further assumption even put a rough temporal–frequency scale on as well.

In a conventional Fourier series the amplitude represents energy per unit wavenumber,

$$F(x) = \int_{-\infty}^{+\infty} f(k)\, e^{ikx}\, dk \tag{1.1}$$

which defines the Fourier coefficient f, where,

$$\frac{1}{2}\overline{F^2(x)} = \int_{-\infty}^{+\infty} \frac{1}{2} f^2(k)\, dk = \int_{-\infty}^{+\infty} \frac{1}{2}\, k f^2(k)\, \frac{dk}{k} \tag{1.2}$$

defines the contribution of each wavenumber to the energy; Parseval's theorem about the multiplication of Fourier series is relevant. The first form of equation 1.2 makes the area under the curve proportional to energy if k is the ordinate, the second form if log k is the ordinate. This amounts to the first form being the energy within unit wavenumber of the wavenumber of interest; the second the energy within wavenumber k of the wavenumber of interest.

For many scales of motion the specific kinetic energy $\frac{1}{2}\rho v^2$ is fairly independent of height so this, normalised to density at sea level, is what is plotted in Figure 1.1. We see a broad peak in kinetic energy centred on global wavenumber 6, corresponding to a wavelength of some 6000 km, or wavenumber 10^{-6} m^{-1}, or period of one week. The energy in wavenumbers 1 to 3 is mostly in slow-moving waves whose dynamics is largely governed by the variation of Coriolis parameter with latitude. These we will refer to as 'very long waves', and are idealised as Rossby waves. Energy in the range 3 to 6 to 12 is associated with travelling disturbances which we identify with the 'long waves' and 'cyclone waves' as Eady called them, or depressions or weather systems, of middle latitudes. The small peak at wavenumber 1/100 km, we associate with the transverse scale of fronts, the peak at 1/1 km with cumulus convection, and the tail to smaller scales with the transfer of energy to shorter wavelengths.

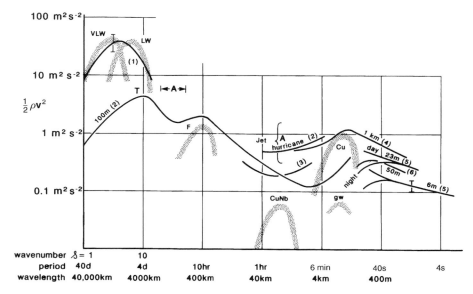

Figure 1.1 Variation of the kinetic energy with scale for tropospheric motion. The wavelength scale is logarithmic, and is labelled with wavelength and with wavenumber, the former being more usual for smaller scales of motion, the latter for larger. A rough temporal-frequency scale is indicated using an advection speed of 10 m s^{-1}. There is one value, marked T, for the tropical experiment of Professor Sheppard, and anomalous values in curly brackets taken in a mid-latitude jet stream. The fuzzy curves represent theoretical expectations due to very long waves, long waves, fronts, cumulonimbus, cumulus, and gravity waves.

Most of the data (see references) comes from dense groups of observing stations set up in middle latitudes in the 1950s and 1960s. There are reasons for thinking that large-scale motion in the tropics is similar in character to that in middle latitudes and the values T and A may represent the long wave and frontal scale for those regions. Curve (2) is from a time-series at 100 m so the horizontal kinetic energy is, because of friction, less than that typical of the large-scale analyses, but with fronts better defined near the ground maybe this part is more reliable. Curve (3) is from a time series taken specifically to identify the meso-scale minimum. Region marked Jet are for turbulent conditions just below the jet stream, and are about the shortest scale observed for the free atmosphere. Curves (4, 5 and 6) show convergence towards a viscous dissipation scale near the ground.

The fuzzy curves show the contributions to the spectrum from 'well defined' motion systems, like very long waves, long waves, fronts, cumulo-nimbus, cumulus and gravity waves, whose appearance is supported by theoretical considerations treated later. We notice that the lines in this 'theoretical' spectrum are much better defined than in the real one. Figure 1.2 shows the corresponding spectrum for vertical motion. Notice that the units are four orders of magnitude smaller than for the horizontal components. How would we expect the two curves to compare ? Well maybe motion up to the scale of the depth of the troposphere (10 km) would be similar in all directions, while motion of larger scale would be inhibited by this depth. The first is true except that the motion becomes inhibited in the vertical for scales of some few kilometres rather than ten. The second is true except that the absolute values of vertical velocity are smaller by a factor of about 100. We will find that it is the vertical stratification of the atmosphere which is responsible for these features. We would like to describe, in general mechanistic terms at least, the shape of this spectrum. We are astonished to see no evidence of cumulonimbus storms, which we expected to find at a wavenumber of about $1/20$ km. We notice, from Figure 1.2, that the cumulonimbus scale makes a more significant contribution to the spectrum of the vertical component of velocity. It is this spectrum that attracts our attention when we seek to describe vertical transfer.

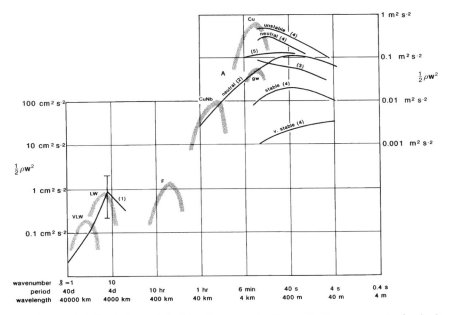

Figure 1.2 Variation of the kinetic energy in the vertical component of velocity with horizontal wavelength. This is nearly the opposite variation with wavenumber to the horizontal kinetic energy; more constrained by the closeness of the ground, specially for longer wavelengths.

1.3 How isolated phenomena appear in wavenumber-space

It is not easy to separate a phenomenon from its environment. The phenomenon is there because the environment is peculiar. The environment is being changed by the action of the phenomenon. Figure 1.3 shows cartoons of a variable in four spatially isolated phenomena, together with their energy spectra. Two types of spectrum have been plotted; the upper one with wavenumber k as ordinate, the lower one with log k as ordinate. We have also chosen to plot amplitude on a logarithmic scale in the lower picture. By so doing we have spoiled the energy matching to area, but not much. All these are arbitary choices, chosen to make some propositions more palatable.

The different events each have halfwidth a, amplitude A, and their spectra do not differ greatly, particularly in the log–log picture. However we do not like the

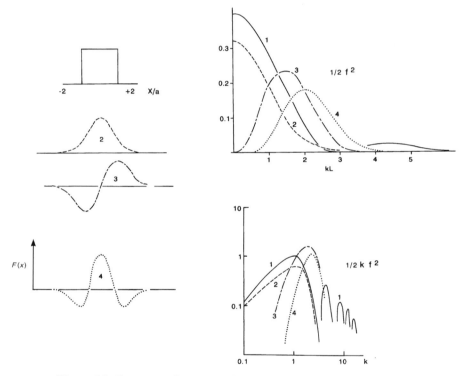

Figure 1.3 Cartoons of some spatial functions $F(x)$ with their Fourier transforms $f(x)$, as in equation 1.1. Events 1 and 2 have much more energy at long wavelengths than the others. When we subtract a compensating downdraught from 1 such that $a'A' = aA$ then the long-wave side of 1 becomes similar to 3 and 4, and similarly for event 2. On the short wavelength side, curve 1 falls off much less steeply (like k^{-1}) than the others as a consequence of the discontinuities at the edges. Event 2 is a Gaussian function in space whose Fourier transform can be written down algabraically by 'completing the square' in the integrand, and 3 and 4 are just the spatial derivatives of this, normalised a bit. We take the composite between 2 and 3 as standard for converting observed events into the spectrum. We are a little worried because the energy in the transforms of such visually different flow regimes as 2 and 3 *are* so similar; the spectrum is not very discriminating over such niceties.

broad spectrum that results when the phenomenon is of one sign, as in 1 and 2. We would prefer to have a small half-width to the phenomenon, identifying more closely a space scale. If we suppose our description of the cumulus cloud to include the slow cloudfree downdraught, even though not as spectacular as the towering cloudy updraught, then we ensure compensatory mass flux, and get something much more like the long-wave end of 3 and 4. Neither do we like the energy in short wavelengths in 1, which decreases like k^{-1} and comes from the discontinuity at the edges. It is physically difficult to generate discontinuities, and where we do invent them as a convenient form of description, they often disguise fundamental deficiencies. We make a special study of such discontinuities in Section 6.1. If we remove the discontinuity by adding a transition zone where the variable falls linearly to zero then the energy falls off like k^{-3} and becomes very similar to the other curves. We argue that events 3 and 4, which contain only deviations and are smooth are 'nicer' than the others and prefer to see isolated phenomena look like them.

The profile 2 is a Gaussian shape, whose Fourier transform can be written down without complexity. The much nicer event profile, number 3, comes from differentiating the Gaussian function, and its transform, and rescaling. We now have a universal translation of isolated phenomenon into wavenumber space. Notice that the spectra of 3 and of 4 are very similar even though the events have a different shape. This is because our spectrum has concentrated on the amplitude, rather than the phase of the motion.

1.4 Repeated phenomena

A second characteristic that we would like to visualise is that phenomena, as well as having a scale of local extent also have one of separation, as with cloud streets separated by clear spaces. If there were just two events separated by distance $2L$, then the energy spectrum would be modulated by $\cos^2 kL$ so there would appear a structure of spacing $\delta k = \pi/L$; a fine structure relative to the scale a^{-1} of the relatively smooth phenomenon. This shows that we expect to see in wavenumber space, the broad band of width $1/a$, given by the continuous spectrum of the local event, as made up of a fine structure of wavenumber-width π/L given by the Fourier decomposition of the spacing. Observed spectra often show fine structure but this is usually smoothed out as representing sampling error. Perhaps we should not be so hasty, perhaps it is not noise, but music. As a final approach to realism we might speculate that for a patch of patterns of width $L^* >> L$ the discrete peaks would have width $1/L^*$ as they began to resemble more closely a Fourier series. Now we are in a better position to superimpose some realistic phenomena on the spectrum of Figure 1.1.

1.5 Contribution of phenomena to the global kinetic energy spectrum

1.5.1 Cumulus

We put $2a = 1$ km, to represent the width, $A = 5$ m s^{-1} for the vertical component of velocity and 2 m s^{-1} for each horizontal component, $L = 10$ km, to represent the separation. The energy peak of 2 m^2 s^{-2} is at $k^{-1} = 220$ m. The corresponding spectral distribution can be seen in Figures 1.1, and 1.2 . The goodness of fit strengthens our supposition that this part of the spectrum is largely represented by the physical process of cumulus convection.

1.5.2 Cumulonimbus

With three components of wind (relative to the storm) of 5 m s^{-1}, and updraughts of width 10 km, separated by 3000 km (as the distance between cold fronts) gives maximum kinetic energy of 0.1 m^2 s^{-2} near $k^{-1} = 3$ km. We verify that such severe storms make an insignificant contribution to the spectrum of total kinetic energy as in Figure 1.1. We wonder what parameter we ought to invent in order to measure the undoubted destructive power of such storms. We notice that our analysis largely excludes catastrophes and wonder why that is.

1.5.3 Frontal zones

$L = 3000$ km, $a = 100$ km, 7 m s^{-1} of only one (the tangential) component of horizontal wind, gives 1.2 m^{2}s^{-2} , at $k^{-1} = 60$ km. There is some small doubt that this bump is resolved by the data, but I thought it should be there so used the one bit of data that showed it.

1.5.4 Long waves and cyclone waves

Values of $2a = 3000$ km, L $= 6000$ km, 10 m s^{-1} of each of two horizontal components of wind, contribute 20 m^2 s^{-2} at $k^{-1} = 900$ km. We notice that the wavelength, rather than reciprocal-wavenumber, becomes a more tangible representative of scale as the phenomenon begins to occupy more of the available space.

1.5.5 Very long waves

Values of $2a = L = 10\,000$ km, 5 m s^{-1} of horizontal velocity, contributes 20 m^2 s^{-2} at a wavelength of 18 000 km.

1.6 What do we learn ?

The main gain is the forced rationalisation, and the change in viewpoint demanded by the conversion from phenomenological to spectral data. That a high frequency "tail" must be added to all our event-spectra implies that the real atmosphere is more diffuse than our event-models. This tail is usually described in terms of a cascade of energy to shorter wavelengths, with molecular viscosity as the ultimate sink for kinetic energy. I find it difficult to believe that energy cascades all the way from the 1000 km scale of weather systems to the mm scale of molecular viscous dissipation, but we shall see.

1.7 Good and bad descriptions of phenomena

We are reluctant to name anything that does not last long. We notice that a cumulus cloud lasts rather a long time especially when we take care to follow it across the sky. Frank Ludlam thought he had sometimes seen a cumulus cloud reappear where it would have been had it not disappeared in the meantime. This certainly happens with a wave moving through a simple wave packet, because it spends half its life with negative amplitude, so why not with a cumulus ? It is often more sensible to take axes moving with the phenomenon rather than relative to the surface of the earth. Describing the action of a motor car engine as seen by a pedestrian would seem perverse, and raise wonder in him that the different bits all happened to keep pace with each other, with tiny but crucial, differences in the velocity of the valves and the ports they were to uncover. But much descriptive meteorology is done precisely that way.

The weather patterns of middle latitudes are thought to arise as a consequence of instability of the mean flow, and to represent the main mechanism for exchanging parcels of warm moist air from the tropics with cold dry ones from the polar regions; thus contributing to the global scale balance of water vapour and thermal energy. The weather pattern moves at a velocity intermediate between that of the parcels it is engaged in exchanging. Figure 1.4(a) shows a pattern of sea-level isobars (or streamlines) typical of such systems, while Figure 1.4(c) shows the 700 mb flow but now relative to the system. This is easily obtained by adding on a pressure field for the motion of the pattern. The first picture suggests a whirlpool indeed a cyclone of flow in which air goes round, while the second suggests a sort of shovel, or conveyor belt, steadily carrying potentially warm air polewards and potentially cold air equatorwards. We notice how much more elongated in the transverse direction is the relative pattern.

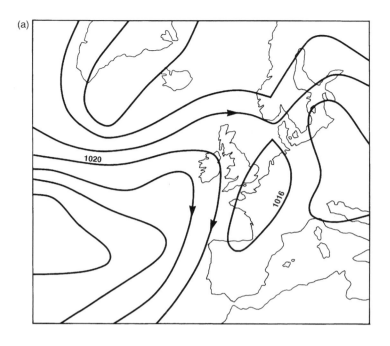

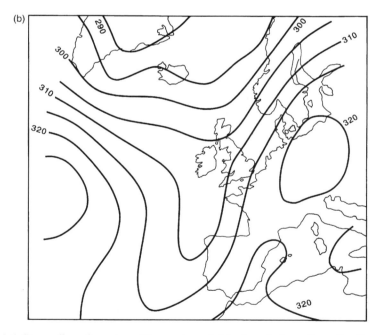

Figure 1.4 Streamlines for a travelling system. (a) Isobars (streamlines for the geostrophic flow) of the surface flow for 9 July 1959, when a depression was crossing southern England. This is the picture we are usually given as the context of current weather. (b) Height of the 700 mb surface showing geostrophic streamlines for the flow at that level. This is about the steering level for the wave so we imagine the flow to be more typical of the bulk of the system. (c) Flow relative to the moving system at 700 mb. There is no depression in the sense of closed streamlines, and air certainly does not seem to go round anything. Really much more like a trough, or a wave. The curvature of the paths of the air is not the same as the curvature in the surface isobars, indeed

(c)

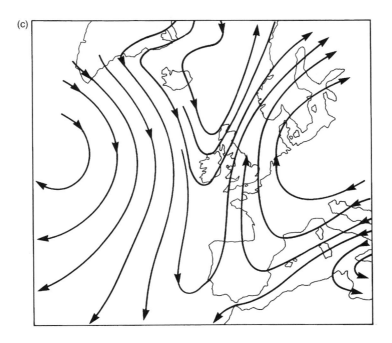

(d)

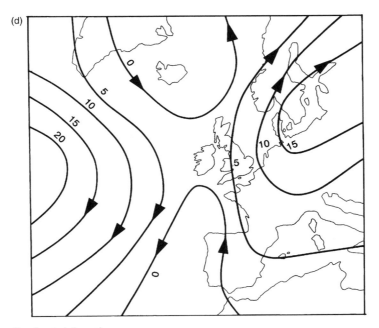

Caption for fig. 1.4 (*cont.*)
sometimes it is of the opposite sign. We now visualise the flow as an exchange between the warm air to the east and the cold air to the west. (d) Relative flow along a typical isentropic surface. This is as close to a trajectory picture as we can get without going to a detailed time-dependent isentropic flow. Incidentally, if we did this we would have the occasional clash where two parcels of air would be at the same place, though at different times which is very confusing to the eye. As we pass through this series of pictures we see more clearly the processes active in determining the flow. The sequence also shows that the transverse displacements become more obvious as the more realistic representation is taken.

Even more spectacular is the impression given by the isentropic relative-flow picture of Figure 1.4(d). There we notice from what very low latitudes the warm air comes. We may become aware of this when dust from the Sahara dessert is rained out in Europe. Figure 1.5 shows relative flow for a particular spectacular storm, again very much a shovel taking warm air poleward.

Cumulonimbus clouds also tend to move with a velocity typical of the middle-level winds and to occur in strongly sheared flow. Thus they overtake the surface air and are overtaken by the air at upper levels. The air at ground-level is potentially warm, largely because of the massive store of latent energy associated with its high water vapour content. The upper air is dry and potentially cold. This potential can be realised and the upper air brought down with downward buoyancy if water can be evaporated into it.

Figure 1.6(c) shows a flow pattern that seems as if it can achieve this. The warm air slopes up over the cold allowing liquid water to fall out of the warm into the potentially cold air. This makes for a rather beautiful efficient circulation. Unfortunately, it is not consistent with dynamical considerations. We found this out while trying to solve for the free streamline separating the two flows, which had the opposite orientation of warm over cold. It can be described a little more simply. Thus we notice that the warm air would be moving fastest at the top of its path where the positive buoyancy has been realised, and will therefore be reluctant to turn the sharp corner at the top. Similar arguments apply to the cold air, which finds the sharp corner at the bottom difficult to negotiate because it would be travelling fastest there. These arguments can be elaborated by application of Bernoullis' equation along the dividing streamline. This then shows that the flow is likely to take the opposite orientation, as shown in Figure 1.6(d). But now the water condensed in the warm moist air cannot fall into the dry potentially cold air, and we have spoiled the beautiful system.

Figure 1.6(e) shows that by using the third dimension we can still bring the potentially cold air under the rain. Our storm has now (realistically) gained parity, in the sense that the updraught and downdraught intertwine, either to the right or to the left and has become properly three dimensional.

Cumulus are difficult. They do not seem to last very much longer than the time it takes for air to pass through them; whether they reappear again later, or not. Ludlam pointed out that by the time you have flown up to an incipient cumulus it is already at a rather mature stage in its lifetime. Their motion patterns are so weak as to defy easy identification. Time-lapse pictures suggest that there is often a preferred source at low levels where clouds grow and that these clouds behave as rather inefficient cumulonimbus as they move away. Thus while most of the downdraught may be in the form of slowly descending clear air, there may also be something like a gust front, where air cooled by evaporation reaches the ground, and there peels off potentially warm air. Such action may be inferred from the characteristic sawtooth pattern to the variation of surface temperature with time. The source region for cumulus may be

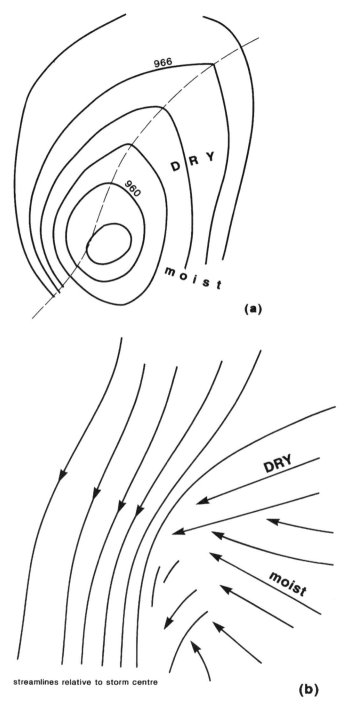

Figure 1.5 Flow in the great storm of 6 October 1987 over southern England. After development that was explosive in some senses the storm spent about 18 hours in almost steady state as it swept across southern England. (a) Isobars at 03 Z. (b) Streamlines relative to the storm covering 00 Z to 06 Z.

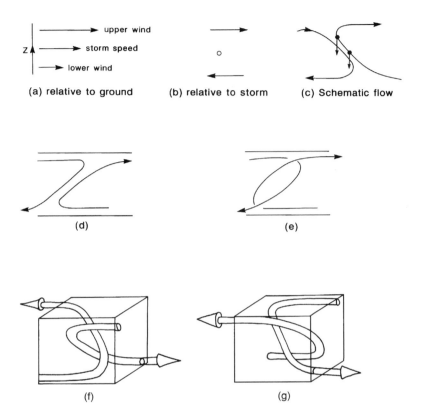

Figure 1.6 Motion relative to the system for several models of cumulonimbus convection which are idealisations for convection in sheared flow. A storm embedded in the highly sheared flow (a) relative to the ground compromises on the disruptive effect of the shear by adopting an intermediate propagation velocity giving apparent flow (b), which is consistent with flow (c) which allows liquid water to fall out of the updraught, and into the downdraught, both of which serve to accelerate the flow. However, application of Bernoulli's equation, or visualising the forces needed to guide parcels round corners shows that (d) is the only dynamically consistent solution. Trajectories (e) reconcile this orientation with precipitation falling from updraught to downdraught giving realistic parity and finite width to severe storms as in (f) and (g).

topographic, like a warm lake, or it may be some aspect of advection of air by a larger-scale system like a sea breeze.

Convective activity in the tropics is often irregular on a timescale of a day or two. There is some evidence that this activity moves from E to W associated with the activity of easterly (moving) waves. These are rather poorly defined by the standard observation network, but may be seen by the willing eye, on satellite pictures of the tropical North Atlantic. The action of this wave may be to maintain convective activity by removing the descending warm air before it stifles the convection. There is some evidence that the tropical convection is particularly active when the low-latitude wave intersects a mid-latitude wave, perhaps ensuring especially efficient transfer of warmed air away from low latitudes.

This mixture of wavelike propagation, with a phase–group relation between environment and phenomenon, and with wave–wave interactions, are the sort of concepts that a casual study of the observations suggests. We will explore mathematical models of idealised flows to help us ask more telling questions of the observations.

Chapter 2

Notation

2.1 What we mean by notation

It is useful to clarify some concepts by thinking carefully about what we mean by rate-of-change with time, and by arranging a suitable notation. The partial derivative $\partial/\partial t$ is used to denote the rate-of-change with time of a function of several independent variables when all the variables except t are held fixed. Strictly, the symbol requires a note of exactly which variables are being held fixed, as is done in classical thermodynamics for example. Generally we are able to use the convention that the variables being held constant are as specified in our original coordinate system. Remember, however, that in this case, $\partial/\partial t$ keeping horizontal position and height or pressure or potential temperature fixed, are all different quantities, even though they may appear as the same symbol. Sometimes it is convenient to have the axes moving: following the rotation of the earth, or a particular cloud for example.

2.2 The substantial derivative

The general derivative for arbitrary small changes in the spatial coordinates x, y, z and time t is;

$$\text{Lim}_{\delta t \to 0} \frac{\delta Q}{\delta t} = \frac{\partial Q}{\partial t} + \frac{\delta x}{\delta t}\frac{\partial Q}{\partial x} + \frac{\delta y}{\delta t}\frac{\partial Q}{\partial y} + \frac{\delta z}{\delta t}\frac{\partial Q}{\partial z} \tag{2.1}$$

Many physical laws are conveniently expressed as the change in some property of a particular particle. Thus a particle becomes warmer because heat has been added to it, or acquires momentum through the action of a force. It follows that an important derivative is that following a particle. If the particle has velocity components u, v, w, in the coordinate directions x, y, z, then setting

$$\delta x = u\delta t \qquad \delta y = v\delta t \qquad \delta z = w\delta t \tag{2.2}$$

makes the general derivative follow the particle. We, following L. F. Richardson and others, reserve the notation D/Dt for the resulting expression. Thus

$$\frac{D}{Dt} = \frac{\partial}{\partial t} + u\frac{\partial}{\partial x} + v\frac{\partial}{\partial y} + w\frac{\partial}{\partial z} = \frac{\partial}{\partial t} + \mathbf{v}.\nabla \tag{2.3}$$

Substantial derivatives are quite subtle and many of the mathematical difficulties encountered when treating meteorological systems arise from them. Consider two simple examples of physical interpretation.

First visualise the continuous medium as represented by an assembly of buses. The substantial derivative is then the quantity observed by a bus driver who wishes to monitor the number of passengers on his bus. The partial derivative is the similar number noted by the inspector of buses who monitors the operation of the system from a fixed position. We can construct scenarios in which one derivative vanishes while the other does not.

As a more realistic example of the curious nature of the substantial derivative, consider two-dimensional flow of a liquid towards a line sink. Mass continuity demands that the radial component of velocity v be of the form $v = s/r$ with s a constant representing the magnitude of the sink. The radial component of the momentum equation allows us to calculate the pressure field as $p = const. - \rho s^2/2r^2$. We notice that the expression contains s^2. While we can readily visualise that the low pressure at the centre sucks fluid towards the line sink, it is perhaps surprising that changing the sign of s, thereby converting the sink into a source, leaves the pressure field unchanged, and still low at the centre. Pouring water into a shallow container assures us of the plausibility of this result. To see the reason for this, concentrate attention on a particular parcel being driven outward. As Figure 2.1 shows, with a central source, since the outward directed speed decreases with radial distance, the parcel slows down, and the acceleration of the parcel is directed towards the centre. The similar argument shows that the acceleration of a parcel on its way *towards* the sink is again towards the centre. Thus the substantial acceleration is of the same sense in spite of the reversal of the velocity. I am surprised by this result, and expected the acceleration to change sign when we changed the sign of the velocity – even though I am perfectly familiar with contrary cases, like centrifugal acceleration.

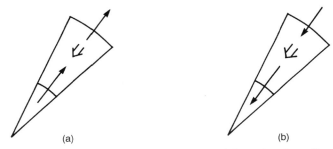

Figure 2.1 Diagram illustrating the substantial acceleration of a particle diverging from a source (a), and converging towards a sink (b). We notice that the substantial acceleration, given by the double arrows, is in the same direction in both cases. While that is apparently alright, what happens if we reverse the argument and start with the pressure field and then let the system go. Does it know which way to go?

2.3 The ordinary derivative

Mathematicians reserve the symbol d/dt for differentiation of a function of only one variable. There may be many parameters but only one variable. Thus we should use the ordinary derivative to define the vertical velocity of a ball dz/dt as the rate of change of its height z with time. The identity of the ball may remain relatively unspecified; Jane's ball, or the one with given position as in the specified problem.

This notation is useful when we need to refer to the rate of change of a property of other groups of particles. Thus we might wish to refer to the total kinetic energy in a system E say, and how that was related to the forces acting within, and upon, the system. Thus we want to write; $E = \int_V \frac{1}{2}\rho \mathbf{v}^2 \, dV_{ol}$ and consider dE/dt to discover for example, that the internal pressure field redistributes kinetic energy between particles, but does not create it.

The line integral of the velocity of a group of particles in a continuum is another interesting property, known as the circulation, related to angular momentum, and to the vorticity. We define the circulation $C = \oint \mathbf{v}.d\mathbf{s}$ and examine dC/dt, to find again that pressure is ineffective. In each of these two cases, the quantity of concern is a function only of the variable time, even though it depends on the initial choice of the set of particles. Neither of the other two temporal derivatives is appropriate, the particles move in space, and consist of an assembly of particles with different behaviour.

2.4 Remaining confusion

Unfortunately, some writers use d/dt for D/Dt as defined here, and some use $\partial/\partial t$ for d/dt. While it is usually clear from the context which derivative they intend, we should insist on reminding ourselves which concept is required.

We have not exhausted the variety of derivatives. Perhaps the most fundamental one remaining occurs in the thermodynamic equation; $\delta Q = dE + p\, dV_{ol}$
equating the change in total energy δQ with the change in internal energy dE
plus the work done on the surroundings by changing the volume $p\ dV_{ol}$. The
change in energy cannot be written as dQ without making $p\, dV_{ol}$ into a total
derivative too, and rendering the system always reversible, which it is not.

Yet other derivatives arise when we follow a system, which might be a
physical entity like a cloud, or a more mathematical one like a wave packet.

Let us write down some simple physical relationships, exploiting our notation and exploring their use.

2.5 The hydrostatic equation

As a relation between pressure and height at a fixed time and horizontal position the supposition that the vertical acceleration of the air can be ignored
compared with g, is quite good; casual observation verifies that air does not
move round like bricks. It may be a misleading approximation depending on
what we want to do with it, but we will come to that later in Section 4.10. To the
extent that the hydrostatic approximation is true, then the vertical component
of the force due to the pressure must balance the weight of the air above:

$$(\partial p/\partial z)_{xyt} = -\rho g \tag{2.4}$$

The density ρ is not usually observed directly and is not so easy to discuss as the
temperature, so it is convenient to eliminate it. Assuming that air behaves like a
perfect gas with constant R, $p = R\rho T$, is accurate enough for our present purpose, the corrections due mainly to the variation in the water vapour content of
the air being negligible in the present context. We then have

$$\frac{1}{p}\frac{\partial p}{\partial z} = -\frac{g}{RT}$$

whence   (2.5)

$$\log p = -\int_{z'=0}^{z'=z} \frac{g}{RT(z')}\, dz' + constant$$

Notice however that we have integrated a partial derivative so the *constant* can
be a function of any of the remaining variables x, y, t. We realise that it is just
the log of the pressure at $z = 0$. Knowing this, and the temperature as a function
of height above, the equation can be integrated to give the pressure as a function
of height, though it is usually used the other way round: given temperature as a
function of pressure from a radio-sounding ascent, to calculate the height as a
function of pressure using equation 2.9, for example.

Up to heights of about 120 km the temperature varies between about 220 and 300 K. Thus the absolute temperature T is roughly independent of height in the sense that z varies much more than T does, and the pressure is given roughly by

$$p = p_0(x, y, t) \exp -(gz/RT_*) \tag{2.6}$$

where T_* is a suitable mean value. Accurately we have

$$\frac{1}{T_*} = \frac{1}{z} \int_0^z \frac{1}{T(z')} \, \mathrm{d}z' \tag{2.7}$$

That the average is harmonic is not unimportant; cold layers act as traps keeping the bulk of the atmosphere closer to the ground than the corresponding isothermal atmosphere and inhibiting its escape to space. Because vapour pressure decreases so rapidly with decreasing temperature, cold layers are even more effective for confining water vapour. Putting in suitable numbers, we find an exponential scale height for pressure, and density, of some 7.5 km, or a decimal scale height of some 16 km. Thus the pressure at 16 km is about 100 mb, that at 32 km 10 mb and so on. This is a useful reminder of the rough height of pressure surfaces.

The mean free path of molecules is inversely proportional to the density so increases similarly as height increases. Thus for a phenomenon of given time and space scales, the kinematic diffusivity becomes more important as height increases, culminating in the steep temperature increase in the thermosphere, and molecular dissipation of tidal and other gravity-wave energy in the lower thermosphere. At about the same height the different behaviour of molecules of different molecular weight becomes apparent, and leads to the diffusive separation of molecules of different molecular weights and affects the rate of escape of molecules from the upper limit of the atmosphere.

At heights greater than about 120 km, there are not enough collisions to ensure local thermodynamic equilibrium. Thus molecules may absorb and emit radiation before they have had time to share their energy in a collision with another molecule. Under these conditions, observation of the effect of drag on satellite orbits gives the density then the hydrostatic equation gives the gaskinetic temperature which will not be the same as the radiative temperature.

2.6 Pressure as vertical coordinate

Instead of using height as vertical coordinate we can use pressure. This has advantages when thermodynamic calculations are needed. By thinking of the physics involved, or by careful manipulation of the partial derivatives involved, we see that

$$\left(\frac{\partial p}{\partial z}\right) = -\rho g \text{ gives } \left(\frac{\partial z}{\partial p}\right) = -\frac{1}{\rho g} \qquad (2.8)$$

Again we can eliminate the density in favour of the temperature and integrate to give

$$z = z_0(x, y, t) + \int_p^{p_0} \frac{RT}{g} \frac{\mathrm{d}p}{p} \qquad (2.9)$$

where p_0 is the pressure at the ground and z_0 is the height of the ground. Should we wish to express the horizontal gradient of the pressure field in terms of the variation in the height of the isobaric surface then Figure 2.2, or juggling with the equations for partial derivatives shows

$$\left(\frac{\partial p}{\partial x}\right)_z = \frac{p(x + \delta x) - p(x)}{\delta x} = \frac{p(x) + \rho g\,\delta x - p(x)}{\delta x} = \rho g\left(\frac{\delta z}{\delta x}\right)_p \qquad (2.10)$$

Similar treatment of the other coordinate gives

$$\frac{1}{\rho}\nabla_z p = g\,\nabla_p z \qquad (2.11)$$

One advantage of the isobaric system is that the same horizontal gradient of the height of an isobaric surface gives the same acceleration at all heights, whereas the same pressure gradient does not. An annoying feature is that the dynamical equations still refer to the horizontal and vertical components of velocity, not the components parallel and normal to the isobaric surface.

A more substantial disadvantage of the pressure coordinate is that the bottom of the system, where boundary conditions need to be specified, is not at a given value of the pressure, and that this value may change with time as well as

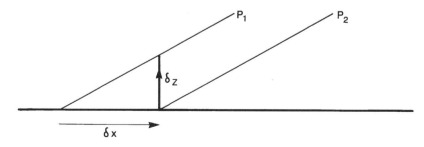

Figure 2.2 Vertical section showing surfaces of constant pressure intersecting the horizontal and the transformation between isobaric and constant height coordinates. A pretty mindless figure until you recollect that it replaces a full general derivative expressed in partial derivatives, constrained to lie along isobaric surfaces. Resolving forces along the almost horizontal isobaric surface we notice that there is no force due to the pressure, but there is now an acceleration due to the component of g in this direction.

space. This is true even when the lower surface is flat. It is necessary that the boundary conditions, and their numerical implementation, do not do substantial spurious work on the system or allow spurious air to enter it. We notice, for example, that the pressure difference between the upwind and downwind slope of a mountain that represents the drag, and is important, is small compared with the difference in the pressure between the top and bottom of the mountain, which is not.

Similar difficulties arise at upper levels. Energy propagates fairly evenly with respect to height but not with respect to pressure. Thus the peculiar physics of the higher atmosphere, like the influence of the middle stratosphere on the troposphere, is really much further away than the 100 mb of pressure difference suggests. It used to be thought that, since the pressure system terminating in $p = 0$ included all the atmosphere it must be better. But the corollary is that the boundary condition at the 'top' must still be very precise and unambiguous, because $p = 0$ is still a long way above the next pressure level.

It might be argued that the pressure driving the motion must vanish at $p = 0$, which turns out to contain no information at all. Nor is it even a plausible condition for, as we have seen, the acceleration depends on the shape of the isobaric surface not on the pressure, so there is a lot of difference between the pressure going to zero, and the pressure going to zero faster than the density. The proposition that the isobaric surface becomes horizontal as $p \to 0$ has enough information to be a boundary condition, but this proposition is arbitrary, unless supported by a proper physical description. One physically satisfying condition is to make the upper boundary a rigid lid, which restricts the velocity field at the upper level. This is physically unrealistic, but physically possible. Another very civilised boundary condition is to make the upper limit a free surface, just like the one we would put at the surface of a heavy liquid. This is just as messy and as physically restrictive as the equivalent condition of an impervious lid at a specified height. There will be more discussion on the nature of boundary conditions in Section 4.4.

Finally, many atmospheric variables like the mean temperature, just do not look right when plotted against pressure. Almost certainly the best compromise is to use $\log p$ as the vertical ordinate for dynamical variables.

2.7 Other coordinates

Where orography is important, $\sigma = \log(p_0/p)$ has the advantage of following the ground, and of being height-like in vertical distribution. Unfortunately the coordinate surfaces still follow the undulations of the ground even far above, where the motion is unlikely to follow them. Again real orography tends to have a much smaller scale than the model resolution therefore steeper gradients. It is the gradients that contribute to momentum transfer, as anyone who has tried to ride a bicycle with small wheels up a high curb will appreciate. It follows that blocking of flow is more likely in the real atmosphere than in the modelled flow. So-called hybrid coordinates have been proposed, arbitrarily becoming more horizontal as height increases. If the flow near the orography were irrotational there might be some advantage in transforming to an orthogonal system satisfying Laplace's equation. The most appropriate system must surely depend on the nature of the dominant physical processes. Ionisation and photodissociation depend on the atomic and molecular absorption of ultraviolet radiation. For these processes height as such seems unimportant compared with the partial pressure of the photo-active gas. Where extinction of radiation is significant, it may be that intensity of radiation, or some surrogate, like $\exp -k p_{partial}$ might be better. Where chemical reactions are dominant, since these depend on number densities, air density together with the mixing ratios of the individual chemicals may well make a more useful system.

Though an application does not immediately spring to mind one might imagine a situation in which the amount of water vapour might make a useful vertical coordinate.

2.8 General transformation of coordinates

While many different coordinate systems are in use, it is important to consider their advantages. Isentropic coordinates may be very useful because those parcels that change their potential temperature comparatively slowly will remain near that surface. On the other hand we might test an isentropic parcel represented in another coordinate system to see how well it retained its original potential temperature. Each time we ensure that a quantity is algebraically conserved we lose the ability to check that our system does nearly conserve that property.

Systems using orthogonal functions can readily be persuaded to conserve energy, but the spectra that result often have too much energy at short wavelengths compared with observation. To make such systems more realistic then demands that we insert an artificial energy sink at short wavelengths which then destroys the energy-conservation property.

In general, systems run into difficulty where bits of coordinate space get cut off and wander through the system; maybe a hump of cold air trapped near the surface that then carries on a secret life of its own. But maybe the same thing happens in other coordinate systems, and there we just fail to notice the anomaly.

Chapter 3

Fundamental equations

3.1 Momentum

We equate the rate of change of momentum of a small parcel of fluid with the sum of forces acting on it. The parcel under consideration is to be imagined as marked with some dynamically insensible agent. Green's magic massless paint will do. In general, the parcel will move changing its shape and volume in the small time interval we have it under observation. The momentum of a small parcel of fluid, of volume $d\tau$, density ρ, moving with velocity $\mathbf{V}$ relative to inertial coordinates is

$$\rho \mathbf{V} d\tau \tag{3.1}$$

The rate of change of momentum of this parcel of fluid as it moves along i.e. the substantial derivative, is

$$\frac{D}{Dt}(\rho \mathbf{V} d\tau) = \rho d\tau \frac{D\mathbf{V}}{Dt} \tag{3.2}$$

because, while $d\tau$ and ρ may change, their product, being the mass of the parcel, does not.

The forces acting on the parcel, roughly in descending order of size are due to gravity, pressure, viscosity, electromagnetism, electrostatic. We consider only the first three here. We notice that 'Coriolis force' and 'centrifugal force' are not on our list, for they are ficticious.

If ϕ is the gravitational potential, then $-\nabla\phi$ is the acceleration due to gravity, and the gravitational force acting on the parcel is

$$\rho d\tau \nabla\phi \tag{3.3}$$

As well as a mean molecular drift, which we have identified with the fluid velocity $\mathbf{V}$, there is an additional 'random' molecular motion. An impervious test surface would be battered by the molecules arriving randomly and being reflected back. This we identify with the gas pressure p alternatively defined as the force per unit area perpendicular to an area which moves with the fluid. The x-component of the pressure force on opposite sides of a Cartesian box is

$$\delta y\,\delta z\,p(x) - \delta y\,\delta z\,p(x + \delta x) = -\delta x\,\delta y\,\delta z\,\partial p/\partial x \tag{3.4}$$

and similarly for the other faces.

The resultant pressure force on the parcel is

$$-\nabla p \tag{3.5}$$

Suppose that μ is the dynamical coefficient of viscosity, then the force due to viscous transfer of momentum is nearly

$$-\mu\nabla^2\mathbf{V} \tag{3.6}$$

Again it is argued in the kinetic theory of gases, that molecules approach our test surface with a tangential component of momentum. This can be found from the supposition that they spontaneously traverse a mean free path of length l, moving at speed c comparable with the speed of sound, terminated by a collision at which they share certain properties typical of their starting place. Since these additional molecular velocities are relative to the mean velocity of the fluid, equal numbers of molecules must pass in each direction and the net rate of transfer of property Q per unit mass, through the face $\delta y\,\delta z$ which moves with the mean velocity of the fluid, is due only to the correlation between the molecular velocity and the property which can be written as

$$-(n\,l\,\delta x\,\delta y)(m\,l\,\partial Q/\partial z)/(l/c) = -\mu\partial Q/\partial z \tag{3.7}$$

where m is the mass of one molecule, and n is the number density of molecules, so that $\mu \simeq nmlc = \rho lc$. Thus we argue loosely that momentum is diffused through the upper face $\delta x\,\delta y$ into the rectangular parcel at rate $\delta x\,\delta y\,(\mu\,\partial\mathbf{V}/\partial z)_{z+\delta z}$ and through the lower face at rate $-\delta x\,\delta y\,(\mu\,\partial\mathbf{V}/\partial z)_z$. For this pair of faces the rate at which momentum is accumulated is therefore $+\delta\tau\,\partial^2\mathbf{V}/\partial z^2$ and similarly for the y- and the z-faces.

Unfortunately, this simple pictorial argument is not quite accurate for momentum. This we can discover by noticing that it implies a spurious stress in a body that rotates as a solid, for the tangential component of velocity varies linearly with distance in such a system. Avoiding this inconsistency requires the rather sophisticated distinction between distortion, dilation, and rotation to be found in standard textbooks, but gives much the same answer for the stress divergence in the end. The reader might wonder why an incorrect argument is given. Well, it is natural, pictorial, faultable and therefore improvable.

It is sometimes said that the molecular motion is random. If it were truly random there would be no correlation between velocity and property, and therefore no transfer. Whenever something is said to be random, think carefully, for random implies something about a method of choice rather than being an intrinsic property. You can choose randomly in a set of black and white balls, but you cannot have a random colour.

Finally, adding these terms we find the equation for conservation of momentum when the measurements of velocity are made in an inertial frame

$$\rho \, D\mathbf{v}/Dt + \nabla p - \rho \, \nabla \phi = \mu \, \nabla^2 \mathbf{v} \qquad (3.8)$$

3.2 Rotating axes

> Thus though we cannot make our sun stand still, yet we will make him run.

It is convenient to measure positions and velocities relative to the Earth. If we did not, we would have to add a W–E component of velocity of speed $\omega r \cos lat$, of some 300 m s^{-1} to the velocities measured relative to the earth to give the absolute velocities of equation 3.8

The formula for the absolute acceleration of a particle observed from a rotating frame is derived in many textbooks. There is no doubt about the result, so here we content ourselves with a simple, but not very exact derivation that serves to keep the principles to the fore.

Suppose that the coordinate system rotates with constant angular velocity ω. The velocity of a particle relative to inertial coordinates, $\mathbf{v}_{abs}$ say, can then be written as the velocity relative to the coordinates $\mathbf{v}$ say, plus the velocity of the coordinate $\mathbf{r}$ say, as

$$\mathbf{v}_{abs} = \mathbf{v} + \omega \wedge \mathbf{r} \qquad (3.9)$$

If we visualise the equation as representing the effect of operating on the position vector in two different ways, one in an absolute sense D/Dt_a, the other relative to the coordinates D/Dt then we can write $D/Dt_a = D/Dt + \omega \wedge$. Allowing this to operate on $\mathbf{v}_{abs}$ we get the absolute acceleration, expressed in terms of quantities measured relative to the rotating axes:

$$\left(\frac{D\mathbf{v}_{abs}}{Dt_{abs}} \right) = \frac{D\mathbf{v}}{Dt} + 2\omega \wedge \mathbf{v} + \omega \wedge (\omega \wedge \mathbf{r}) \qquad (3.10)$$

Let us look for insight, even from such a simple equation. The last term is the centrifugal acceleration of a particle that does not move relative to the axes. It is not obviously negligible for the Earth's atmosphere in spite of being small compared with the gravitational acceleration. Thus, a particle subject to this, apparently small, tangential acceleration would, if displaced gently from the pole, arrive at the equator moving at some 400 m s^{-1}. This does not happen to penguins, or to polar bears for the 'solid' earth, and the oceans are also subject to the same acceleration and have changed shape (to make the planet slightly oblate) just to cancel the component of acceleration along the surface. Thus there is no horizontal component of this acceleration for the real earth. The terms in $-\nabla\phi$ and in $\omega \wedge (\omega \wedge \mathbf{r})$ can be combined to make 'apparent gravity' $g\mathbf{k}$, where $\mathbf{k}$ is the unit vertical vector. So long as we ensure that this $\mathbf{k}$ is perpendicular to the ground we can ignore the non-sphericity of the planet so far as geometric distortion is concerned. We are not so fortunate with Jupiter.

The relevant non-dimensional number for oblateness is $\omega^2 r/g$. Tilting $\mathbf{g}$ round to cancel the tangential component of $\omega^2 r$, we find that it accounts for only about half of the observed oblateness of the Earth. This is because the gravitational attraction is also tilted round by the oblate Earth. Astonishingly, gravitational acceleration does not pass through the centre of gravity but is tilted round towards the bulge, which exaggerates the distortion. That large bodies watched for a long time behave rather like fluids is a nice idea.

That the value of g varies with latitude is not insignificant but can be taken into account (corrected for) by using the geopotential rather than true height as vertical ordinate.

Expressed relative to axes moving with angular velocity ω the vector momentum equation can be written;

$$\frac{D\mathbf{v}}{Dt} + 2\omega\wedge\mathbf{v} + \frac{1}{\rho}\nabla p + g\mathbf{k} = \frac{\mu}{\rho}\nabla^2 \mathbf{v} \tag{3.11}$$

The term $2\omega \wedge \mathbf{v}$ is the Coriolis acceleration. It is perpendicular to the angular velocity of the coordinate system and to the apparent velocity of the particle. If there were no forces acting on the particle, then the particle would have an acceleration $-2\omega \wedge \mathbf{v}$ relative to the rotating axes, which makes the path of the particle in space a straight line traversed at constant speed. Most simple descriptions of Coriolis acceleration leave us with this notion of deflection of particles, subject to no net force, e.g. golf balls deflected towards the left in the northern hemisphere. But for large-scale motion it is the other way round, for it is the relative acceleration which is small, and the force field which is large, just large enough in fact to make the particle more-or-less unaccelerated in the relative coordinates. Perhaps this is because large-scale motion is concerned ultimately with carrying particles efficiently over large distances relative to the Earth's surface?

3.3 Motion independent of longitude

It is worth examining a very simple example of balanced motion, not devoid of meteorological interest. Suppose that, as in Figure 3.1(a), a parcel of air moves from west to east at speed u relative to the Earth. The angular velocity of the parcel is $\omega + u/r \cos lat$, which has horizontal component of acceleration $2\omega u \sin lat + (u^2/r) \tan lat$. The second term is found to be rather small (astonishingly, even where $lat \sim \pi/2$ but that case needs extra thought), and the first is the Coriolis acceleration. If that ring of air continues to move zonally, as it is observed to do in the mean jet streams, there must be a force, to cancel the horizontal component of this Coriolis acceleration. We presume that this force comes from the pressure field, with higher pressure on the equatorward side of a ring moving from west to east. This is the geostrophic balance illustrated in Figure 3.1(a).

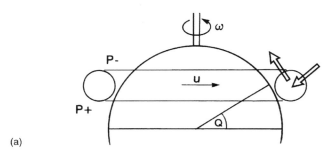

(a)

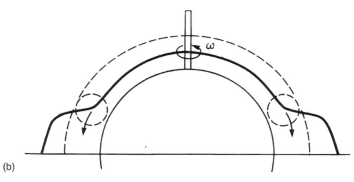

(b)

Figure 3.1 (a) Dynamical equilibrium for a ring of particles moving zonally (west to east) at speed u relative to the surface of a planet rotating at angular speed ω. The weight of the ring is supported by the nearly hydrostatic pressure, which must be greater on the equatorward side to balance the horizontal component of the acceleration. For equilibrium we see that the apparent velocity must be perpendicular to the pressure gradient. (b) The approach to geostrophic equilibrium. If there was no pressure field then the ring would accelerate towards the equator, decreasing its speed relative to the surface of the Earth, and piling up air on the equatorial side. If the rotation is so large that $\omega^2 R^2/gH$ is small compared with about 10, then the ring will acquire the pressure field needed for geostrophic balance. In the Earth's atmosphere it is about 10 so the adjustment on the largest scale between pressure and wind is mutual.

What if there was no such pressure force? In this case Figure 3.1 (b) shows that the ring would start to move towards the equator. As it did so the zonal component of relative velocity would decrease in order to conserve the angular momentum of the ring, and air would pile up on the equatorward side of the ring tending to provide the geostrophic pressure gradient.

In general the ring will achieve rotational balance in the sense of implying small displacements to acquire a typical relative velocity u, if $u/\omega r$ is small. It is indeed small for most planets, and the Sun, but Venus is odd for the solid planet rotates so very slowly that if we use the period of rotation of the planet to give ω then the inequality is not satisfied. However the Venusian atmosphere is in such violent rotation at almost all heights, that most air goes round the planet in only a few (Earth) days, and if we put our axes in that air then the inequality is fairly well satisfied. Is that not curious?

If the whole ring accelerated towards the equator, it would pile up air on its equatorward side, which is in the sense of developing a pressure field to balance the Coriolis acceleration. For an atmosphere of depth H, we find a number like $\omega^2 r^2/gH$ to represent this process. If this number is large, as it is in the Earth's atmosphere, and ocean, and many other atmospheres too, then the pressure field adjusts towards the velocity. We take this to hint that the pressure field may be just as good as the velocity field for describing large-scale motion.

3.4 Continuity of mass

We have already used the law of conservation of mass in deriving the momentum equation, which is quite logical if we remember that forces change momentum, rather than give acceleration. However, this equation for the conservation of mass is in rather an odd form, and may be converted into a more conventional one by the following manipulation. First we differentiate by parts to give

$$\frac{D}{Dt}(\rho\,\delta\tau) = \delta\tau\,\frac{D\rho}{Dt} + \rho\,\frac{D\delta\tau}{Dt} \tag{3.12}$$

but $\delta\tau$ represents the volume of a marked set of particles which may change as they move along. For an initially rectangular volume, as in Figure 3.2 we see that the $\delta y\,\delta z$ face at x moves at speed $u(x)$, diminishing the volume if u is positive, while the opposite face at $x + \delta x$ moves at speed $u(x + \delta x)$, increasing it. Thus the volume increases at rate $\delta\tau\,\partial u/\partial x$, due to the 'pumping' of these faces, and similarly for the other two to give

$$D(\delta\tau)/Dt = (\partial u/\partial x + \partial v/\partial y + \partial w/\partial z)\,\delta\tau$$
$$D\rho/Dt + \operatorname{div}\mathbf{v} = 0 \tag{3.13}$$
$$\partial\rho/\partial t + \operatorname{div}(\rho\mathbf{v}) = 0$$

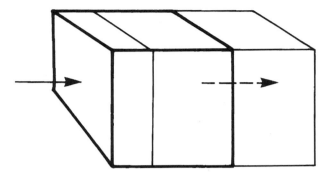

Figure 3.2 Continuity of mass for a marked mass of fluid. Each pair of ends of the Cartesian box move at different normal-velocities so affect the volume of fluid inside. The mass inside is unchanged.

as an alternative expression of the same physics, after a small mathematical manipulation. We see that the second form refers to the change in density at a fixed point in space, as being due to the convergence of mass-flux through the walls of a fixed volume.

For gridpoint models, it is usually an advantage to use this flux form of the equations to ensure that advected properties are not spuriously accumulated in grid-boxes.

The momentum equation can also be derived by treating the momentum budget at a fixed point in space instead of the derivation following a marked parcel. We then need to include terms representing the advection of momentum into the fixed volume, and to use continuity of mass which distinguishes between advection by the fluid of velocity and momentum. The final form reproduces, of course, the substantial form of the equation.

3.5 Continuity of energy

The first law of thermodynamics states that a small amount of energy δQ given to unit mass of a system, which here is our small parcel of fluid, can be used to change the internal energy and to do work against the surroundings. Where the work done by the viscous forces can be neglected the work is communicated only by the pressure force, this can be written

$$\delta Q = \mathrm{d}E + p\,\mathrm{d}V_{ol} \tag{3.14}$$

where V_{ol} is the volume. This is what we previously refered to as $\delta\tau$ and thought of as infinitesimal, but now becomes V_{ol} and is of unit mass whose units are unspecified! The confusion is conventional, and both notations will disappear forever from our analysis quite shortly, so we are not too concerned. Work done by tangential stresses is generally negligible but can lead to inconsistencies in energy budgets where mechanical dissipation by viscous forces is important.

Moist air behaves nearly like a perfect gas except that the latent heat is noticeable. Thus if X is the mass of water vapour in unit mass of air,

$$\mathrm{d}E = C_v \,\mathrm{d}T + L \,\mathrm{d}X \tag{3.15}$$

where C_v is the specific heat at constant volume, and L is the latent heat of evaporation. Eliminating V_{ol} by $\rho V_{ol} = 1$ unit mass, the thermodynamic equation can be rearranged to give

$$\delta Q = C_v \,\mathrm{d}T + L \,\mathrm{d}x + \mathrm{d}(p/\rho) - \mathrm{d}p/\rho = C_p \,\mathrm{d}T + L \,\mathrm{d}x - \mathrm{d}p/\rho \tag{3.16}$$

where we have used the perfect gas equation

$$p = R\rho T \qquad R = C_p - C_v \tag{3.17}$$

Replacing all d's by the substantial derivative D/Dt, and supplying suitable physical descriptions of the diabatic heating Q, and the moisture mixing ratio X gives a useable thermodynamic equation.

3.6 Dry adiabatic

For some uses, and under some circumstances, δQ and $\mathrm{d}X$ can both be neglected. When looking at theory, we can easily and consistently imagine the consequences of having no diabatic heating, and no water vapour. However, when interpreting observations, we often need to take notice of the small difference in density between water vapour and air, and to treat adiabatic motion with great care, noticing the change of specific heat with temperature for example. This because the variations in density that give rise to large dynamical effects are so small compared with the thermodynamic ones. Dividing the resulting equation $C_p \mathrm{d}T - \mathrm{d}p/\rho$ by T we see that it can be integrated to give

$$\frac{D}{Dt}\left(\log T - \frac{R}{C_p} \log p \right) = 0$$

thus
$$\log T - (R/C_p) \log p = \log \theta - (R/C_p) \log p_0 \quad \text{(say)} \tag{3.18}$$

is constant following a parcel in dry adiabatic motion. The quantity θ is of dimensions temperature, with the physical interpretation that it is the temperature that the parcel would attain if brought dry-adiabatically to the reference pressure p_0. Because of this property it is known as the 'dry bulb' potential temperature. It is usual to take p_0 to be 1000 mb, but if one were concerned with, say, motion in the mesosphere $p_0 = 10^{-3}$ mb might be more appropriate.

In practice θ is indeed usefully conserved in the free atmosphere, with δQ giving temperature changes of something like 1 K per day mainly due to electromagnetic radiation, while T might easily change by 30 K in the same time. Whether θ is conserved or not it is usually a more useful expression of the thermodynamic quality of the parcel than, for example, the gas kinetic temperature T, but we still need T if, for example, chemical reactions through the Arrhenius effect, or thermal radiation through the radiative emission, are important.

3.7 Wet adiabatic

Suppose that δQ is negligible, but that the air is saturated with respect to water vapour, and remains so as it moves. This turns out to be a reasonable approximation for the ascent of air in clouds. A fundamental law of thermodynamics is that the saturation vapour pressure is a function of temperature only, i.e. it does not depend on the partial pressure of any additional gas. We can therefore define the saturation vapour pressure of water $e_s(T)$.

Classical thermodynamics shows that e_s satisfies the relation of Clausius–Clapeyron, and varies exponentially with $1/T$, and therefore nearly exponentially with T, roughly doubling for each increase of 10 K in T. This carries with it a picture of molecules in the liquid acquiring enough energy to overcome a potential barrier E, associated with surface tension at the liquid–vapour boundary, to exchange molecules with the vapour. The Boltzman factor $\exp E/kT$ then supplies the form of the relation. The saturation mixing ratio X_s is connected with e_s by the gas law for water vapour. Since water vapour under atmospheric conditions is tolerably close to a perfect gas we can put; with R_v the gas constant for water vapour, $e_s = R_v \rho X_s T$, and deduce $X_s = \epsilon e_s/p$ where $\epsilon = R/R_v$ is the ratio of molecular weights of water vapour to air. Putting this expression for X in the thermodynamic equation gives, for saturated adiabatic variation,

$$\delta Q = 0 = C_p \, \mathrm{d}T + L \, \mathrm{d}X_s - \mathrm{d}p/\rho = C_p \, \mathrm{d}T - \mathrm{d}p/\rho + \epsilon L \, \mathrm{d}(e_s/p) \quad (3.19)$$

According to fundamental thermodynamic principles, on dividing this equation by T, it should again be integrable. As we have seen the first two terms are, so the third should be, but it is not. There is therefore something wrong with the equation, and the theory leading to it. We can juggle the last term, which is the one causing the trouble, by putting

$$\frac{1}{T}\mathrm{d}\left(\frac{e_s}{p}\right) \simeq \mathrm{d}\left(\frac{e_s}{pT}\right) \quad (3.20)$$

This is an an acceptable approximation if $\mathrm{d}e_s/e_s \gg \mathrm{d}T/T$, and, as we noticed above, $\mathrm{d}e_s/e_s$ is about unity for $\mathrm{d}T = 10$ K, so with $T \sim 250$ K the step is well justified.

We then integrate and discover the wet-bulb (or saturation) potential temperature θ_s defined by

$$\log \theta_s = \log T - \frac{R}{C_p} \log \frac{p}{p_0} + \epsilon \frac{L\,e_s(T)}{pT} - \epsilon \frac{L\,e_s(\theta_s)}{p_0\theta_s} \qquad (3.21)$$

The process of realising the wet-bulb potential temperature is a little more complicated than for the dry process. For all its elaboration equation 3.21 merely says that the parcel of air must be brought to the reference pressure p_0 being kept saturated while doing so, to realise the wet-bulb potential temperature. Remarkably, equation 3.21 is rather more accurate than equation 3.19, from which we appear to have derived it. This is what I call inspired empiricism. Serendipity comes in somewhere too. It must be cheating in some way, but it serves to strengthen our belief that nicety is often a guide to improvement. Had we done the thermodynamics properly no such arbitrary adjustments would have been needed. This means taking proper account of the partial pressure of the dry air; $p - e$, and of the partial pressure of the water vapour e separately, then adding the two together with energy conversion at the phase change; all a bit elaborate, but necessary for strict consistency.

3.8 Thermodynamic processes

There are two principal uses for the thermodynamics. We can use it to replace the sensible variable T by the process variable θ or θ_s where appropriate, or we can visualise it as describing an idealised variation of parcel property during hypothetical motion. Used in this sense we notice that our parcel always has two possible paths in the T–p plane, one in which it is saturated, the other in which it is effectively dry. This cannot be strictly true because the dry parcels can have a variety of vapour pressures. Again, what happens to the condensed water in the parcel? Thermodynamically it should be carried with the parcel and subject to the same temperature changes. This is not very realistic.

Cumulus and cumulonimbus convection are processes whereby energy is transferred upwards through the lower troposphere. In the parcel theory, air near the ground receives energy from the sunshine, as sensible and as latent heat, as they warm and moisten. When the density of the parcel has decreased sufficiently, it tends to rise, with its temperature decreasing with height at the dry adiabatic rate, until it becomes saturated at the base of the cloud. Thereafter its temperature decreases at the lesser rate related to the wet adiabatic process, finally coming to rest near an upper level of zero buoyancy a few kilometres above the ground. The liquid water may follow a different path to the air.

A parcel starting at a pressure of 1000 mb and temperature of 10°C would attain a temperature of −73°C at 300 mb (roughly the height of the tropopause) if it ascended along the dry adiabatic, and a temperature of −58°C if it ascended along the wet adiabatic. The difference arises from the condensation of the 7.8 g kg^{-1} of water vapour in the air saturated at 1000 mb and 10°C. We notice that this would give a temperature change of about 18°C if the release of energy took place at 1000 mb. The difference between the two values for the heating due to the condensation of water vapour (18 K and $73-58 = 15$ K) is not due to the approximation to the equation that we have used, nor to the necessity of accounting for the fate of the condensed water, but to the fundamental nature of thermodynamical equilibrium, and reversibility.

If the water were condensed out at 1000 mb, then the air lifted to 300 mb the resulting temperature would be $-60°$C. Lifted to 300 mb along a dry adiabatic, then condensing out the water at 300 mb would give a temperature of $-55°$C. These two, and the wet adiabatic process are all different processes, more-or-less realisable in some circumstances. The isobaric processes are perhaps not very 'realistic' in an atmospheric context. They correspond roughly to a kind of steam engine driven by the temperature difference between the 1000 and 300 mb surfaces, with the cloud playing the role of the piston. Introduction of the ice phase adds complication. The latent heat of fusion is to be added and the behaviour of a mixture of ice and water particles needs to be considered both as regards their different fall speeds and their mutual thermodynamics.

3.9 Energetics

Taking the scalar product of velocity and equation 3.11, using mass continuity equation 3.13 and the thermodynamic equation 3.14, we can deduce an expression for conservation of total energy. For example in a thermally isolated system the sum of fluid-kinetic energy $\frac{1}{2}\rho v^2$, molecular-kinetic enery $\rho C_v T$, latent $L\rho X$, and gravitational-potential energy gz, is constant. This is of course a fundamental law of physics, but it is not a very useful property for understanding the nature of meteorological motion. This is essentially because the energy transformations it expresses are not meteorologically plausible. For instance, conversion of $C_v T$ into kinetic energy of fluid motion would yield wind speeds of order 300 m s^{-1} which is far in excess of those observed. In fact we can think only of rather unmeteorological ways of achieving that transformation, like using a Maxwell-demon cricketer who would deflect the molecular particles all into the same direction, or by exhausting a gas into a vacuum to make available the molecular energy. To transform gz into $\frac{1}{2}\rho v^2$ would be like dropping a brick of air from the tropopause to the ground to release the gravitational potential energy of air at the tropopause. There is no plausible meteorological mechanism for doing any of these things and the energy released in atmospheric motion is, in each case, very much less. To take this question further we need to recognise the balanced nature of atmospheric motion, and will return to study such energetics in Section 5.1 .

3.10 Completeness

Have we written down enough equations to describe a physical system yet? One way of answering this question, and it may be the only way, is to see if we can predict the state of the system some time ahead, at least in principle. Suppose that we know the heat sources and sinks, for example, that the motion is dry and adiabatic, and we know the initial state of the system, by which we mean that we know the values of the variables $\mathbf{v}$, p, ρ, T, all over the domain at $t = 0$. We might then argue that we can calculate all the spatial derivatives needed in the equations 3.11, 3.13, 3.18 and gas, to give $\partial\mathbf{v}/\partial t$, $\partial\rho/\partial t$ $\partial p/\partial t$ at $t = 0$, hence predict all values of the variables a little time ahead. The set therefore seems to be complete at least for dry adiabatic motion.

But then we have now proved too much, for we must need to specify boundary conditions, such as the temperature, shape and velocity of the bounding surfaces, before the physical problem is properly posed. Is the proverbial butterfly about to flap a wing, is the air blowing trees about, or raising waves on the sea for example?

Our argument is indeed faulty where we supposed that all spatial derivatives were defined. This turns out to be true everywhere except at the very edge of the fluid, where there is no more fluid on one side. There the derivative normal to the boundary is not defined by the internal points and must be supplied by consideration of the physical nature of the boundary.

For example, one special case of our equations is that for thermal conduction

$$\frac{\partial T}{\partial t} = k\,\frac{\partial^2 T}{\partial z^2} \quad 0 < z < H \text{ say} \tag{3.22}$$

Our original argument would lead us to suppose that we could predict the temperature along the finite length of the bar, knowing only conditions along the finite length at an initial time. We know that this is incorrect because we can arbitrarily change the temperature at either end using an external agency. We associate the discrepency with the difficulty of calculating $\partial^2 T/\partial z^2$ at the ends. To close the problem properly we must specify the temperature at these points, perhaps varying with time, or the heat flux which defines $\partial T/\partial z$ there.

Another special case of some interest is the equation for propagation of elastic waves

$$\frac{\partial^2 p}{\partial t^2} = c^2\,\frac{\partial^2 p}{\partial z^2} \tag{3.23}$$

Again our original argument is wrong, and we can easily show that the influence of the boundaries propagates inwards at speed c, which at $300\,\mathrm{m\,s^{-1}}$ covers the depth of the troposphere in 30 s. We might ask, what physical property is it that propagates at this speed? We find that it is imbalances of pressure, like the deviation of the pressure from the hydrostatic value. It is like a bomb exploding, when a pressure pulse propagates outwards. Like me leaning against the wall, where I have to wait until the wall bends enough to balance the force of my shoulder. So called 'stiff' equations are appropriate here. It is important to notice such physical requirements, because we can think of various arithmetic devices for estimating the boundary derivatives from interior values; by fitting polynomials to internal points, for example. Our argument shows that such devices are spurious, and are likely to imply peculiar unphysical conditions at the boundary.

Yet another special case represents the propagation of gravity waves. One class is like those found on the free surface of liquids. They serve to change the mass of liquid in a column, and therefore the hydrostatic pressure at the bottom. Another class is like the waves found on the internal interfaces of superposed immiscible liquids of different densities. They serve to propagate the vertical variations in hydrostatic pressure. We examine some of these processes in more detail in the following chapters.

3.11 Vorticity

It is sometimes useful to reduce the number of variables in our equations. The reason for this is not clear to me. When we describe a system we write down simple relationships between relevant variables; in the present case the primitive equations expressing continuity of mass, momentum, heat etc., but then eliminate variables to discover wave and diffusion equations as special cases. For example, the pressure can be eliminated by suitably differentiating the momentum equation. This consists essentially in integrating round a closed circuit so that the effect of the large potential force ∇p, due to the pressure field, vanishes identically. Not surprisingly, this integration round a circuit produces a vector equation for the line integral of the velocity, which is the vorticity. Alternatively we can take the curl of the momentum equation to arrive at the same equation.

Taking the curl of equation 3.11 is messy because the advective part of $D\mathbf{v}/Dt$ is not amenable to manipulation using standard vector formulae. However $\mathbf{v}.\nabla\mathbf{v}$ can be rearranged, using the formula for $\nabla(\mathbf{a}.\mathbf{b})$, to give

$$\frac{\partial \mathbf{v}}{\partial t} + (2\omega + \operatorname{curl}\mathbf{v}) \wedge \mathbf{v} + \nabla(\frac{1}{2}\mathbf{v}^2) + \frac{1}{\rho}\nabla p + g\,\mathbf{k} = \nu\,\nabla^2\mathbf{v} \tag{3.24}$$

This brings to our attention the absolute vorticity $\operatorname{curl}\mathbf{v} + 2\omega = \mathbf{Z}$ (say) and the pressure-like variable $p + \frac{1}{2}\rho\mathbf{v}^2$ the so-called dynamic pressure. For instance we can reconstruct one form of Bernoulli's equation from equation 3.24, and notice that the pressure p that appears in our equations is that measured by a barometer moving with the air; this pressure is a Lagrangian variable, whereas that read by a stationary barometer is more like a dynamical pressure. For our present purpose we can now take the curl of equation 3.24 and simplify, using fairly standard vector formulae. We get

$$\frac{D\mathbf{Z}}{Dt} - (\mathbf{Z}.\nabla)\mathbf{v} + \mathbf{Z}\,\text{div}\,\mathbf{v} - \frac{1}{\rho}\,\nabla\rho \wedge \frac{1}{\rho}\,\nabla p = \nu\,\nabla^2\mathbf{Z} \qquad (3.25)$$

We now see that wherever the vorticity occurs, it is always the absolute vorticity and never that relative to the coordinate system on its own. This reminds us of the general principle that a law of physics must be independent of the coordinate system used for its expression. The vorticity equation 'knows' that there is an absolute system of inertial coordinates – even when we do not use it.

3.12 The terms in the vorticity equation

The first term says that the absolute vorticity of a parcel of air may change as the parcel moves around. The remaining terms can be interpreted in terms of this effect. The second term shows the effect of variations of velocity in the direction defined by the vector vorticity. We can think of two components of velocity; that parallel to the vortex, and that perpendicular to the vortex. Variation of the component of velocity along the direction of the vortex can be thought of as a stretching, or as unstretching, the vortex tube. At first sight this can be interpreted as changing the radius of gyration round the vortex tube so enhancing, for stretching, or diminishing, for unstretching, the tube. Variation of velocity perpendicular to the vorticity vector acts in the sense of tipping the vortex into another direction. The third term vanishes with the 3-D divergence, but we need to consider the divergence field in order to interpret vortex stretching in terms of radius of gyration. We see that the divergence effect and the vortex stretching effect are not unrelated. Consider the two terms resolved in the direction of the vorticity $\mathbf{Z}$, temporarily defining a Cartesian set of coordinates x, y, z with velocity components u, v, w. The contribution to the rate of change of z-vorticity is

$$\mathbf{Z}.\nabla\,V_z - Z_z\,\text{div}\,\mathbf{V} = -Z_z\,(\partial u/\partial x + \partial v/\partial y) \qquad (3.26)$$

so by suitably combining the divergence and stretching terms we see that it *is* the divergence in the plane pependicular to the vortex that changes the vorticity, which can be interpreted as the change in the radius of gyration of the parcel . The fourth term shows that if the pressure field is not parallel to the density field, rotation will be generated, in the sense of turning a conserved density field to be more nearly parallel to the pressure field. This is known as the baroclinic term, in the sense that it depends on the inclination between the pressure and density surfaces. We might notice that the occurrence of this term means that we have not eliminated the pressure, as was our avowed intention. However, for an important class of large-scale motion systems, the hydrostatic aproximation will suffice for this part of the pressure, and we can write $\nabla p = -\rho g\mathbf{k}$ in this term, which then does eliminate the pressure. Finally, on the right side vorticity will be diffused by viscosity. We notice that all terms except these last two say what will happen to vorticity we already have. Only these last two can generate or destroy vorticity. The baroclinic term does so in response to variations in density which give different accelerations for the same pressure gradient, and viscosity which allows vorticity to diffuse into the fluid from non-slippery boundaries.

When the wind blows across the surface of a body of water, we usually notice that, after some time, the water has begun to move, and move faster near the surface where the wind stress is acting than lower down where it is not. While that seems quite natural, it implies that vorticity, represented by the shearing motiom, has been developed in the water. This is more difficult to explain. Viscous diffusion of vorticity as far down in the water as the shear penetrates takes much too long, and has to be aided by other processes. It seems likely that the intense vortex sheet generated close to the surface by viscosity is advected into the bulk of the water where it represents the sub-surface shearing layer.

3.13 Circulation

Taking a line integral of the momentum equation shows the vorticity equation in a different way. We integrate the tangential component of velocity round a specified line of particles; Green's magic massless paint describes a closed circuit in the fluid which then blows along with the flow. We define the (relative) circulation C_r by $C_r = \oint \mathbf{v}.\mathrm{d}\mathbf{r}$ where $\mathrm{d}\mathbf{r}$ is one of the small increments in position vector that together map out the circuit. Then

$$\frac{\mathrm{d}C_r}{\mathrm{d}t} = \oint \frac{D\mathbf{v}}{Dt}.\mathrm{d}\mathbf{r} + \oint \mathbf{v}.\mathrm{d}\mathbf{v} \tag{3.27}$$

but notice that $D(\mathrm{d}\mathbf{r})/Dt = \mathrm{d}\mathbf{v}$, in the second term, which therefore vanishes round a closed circuit, also the proper use of $\mathrm{d}/\mathrm{d}t$ in this description. A similar argument shows that

$$\oint 2\omega \wedge \mathbf{v}.\mathrm{d}\mathbf{r} = \frac{\mathrm{d}}{\mathrm{d}t}\oint 2\omega \wedge \mathbf{r}.\mathrm{d}\mathbf{r} \tag{3.28}$$

which is the vorticity of the fluid moving with the axes, and which reintroduces the absolute circulation C which is otherwise obvious from equation 3.25. Using the dynamical equation 3.11 for $D\mathbf{v}/Dt$, the circulation equation 3.27 becomes

$$\frac{\mathrm{d}C}{\mathrm{d}t} = -\oint \frac{\mathrm{d}p}{\rho} + \nu \oint \nabla^2\mathbf{Z}.\mathrm{d}\mathbf{r} \tag{3.29}$$

This form of the vorticity equation is very pictorial. The left-hand side shows that all the vortex stretching and tipping can be visualised as being a consequence of the vortex lines being embedded in the fluid and moving with it. The first term on the right vanishes identically if the density is a single-valued function of the pressure, and we see that circulation can be generated inside the fluid only if the lines of constant pressure do not coincide with those for constant density, i.e. the fluid is 'baroclinic', or because vorticity is transferred by molecular diffusion.

In many situations both these terms are small in comparison with the component parts of dC/dt and an equation like 3.29 can be used to predict the future behaviour of the fluid. We use the current velocity field to find the position of the circuit a small time later, and the known conserved value of the circulation to construct the new field of velocity, and so on in a step-by-step integration. The algebraic form of the baroclinic effect can be recovered if we express the line integral as $(1/\rho)\nabla p \, . \, d\mathbf{r}$ and use Gauss' theorem to form $\operatorname{curl}(1/\rho)\nabla p \, . \, d\mathbf{S}$ which takes us back to the vorticity equation again.

On a vertical section, we construct a circuit along lines of constant p and the vertical. This shows that the baroclinic effect is merely the same large vertical gradient of pressure acting on parcels of slightly different density to produce different accelerations. We also see that it represents a tendency for vorticity to be generated in the sense of rotating relatively-passive density surfaces towards being more nearly parallel to the pressure surfaces. Whether this actually happens depends on other considerations. For example the thermal wind equation is hidden away in the rate-of-change of the solid body part of the circulation equation 3.29, so one solution is thermal-wind equilibrium between the wind and the fields of pressure and density.

Even in systems close to hydrostatic balance, we may not put $dp = -\rho g \, dz$ to deduce that, because the circuit is complete in height, therefore the baroclinic term vanishes identically.

3.14 The shallow atmosphere

The atmosphere may be considered as shallow to a good approximation, for some purposes. Thus the term $2\omega \wedge \mathbf{v}$ can be decomposed into the usual Cartesian coordinates with z vertical, x west to east as;

$$2\omega_y\,\mathbf{v}_z - 2\omega_z\,\mathbf{v}_y \qquad 2\omega_z\,\mathbf{v}_x \qquad - 2\omega_y\,\mathbf{v}_x \qquad\qquad (3.30)$$

but shallowness demands at least that v_z/v_{hor} is small for motion of large horizontal scale. We then notice that all the terms containing horizontal components of the planetary rotation are neglected, leaving only the rotation about the vertical, ω_z. Moreover the equations behave as if the only rotation was that about the local vertical. The quantity $2\omega_z = 2\,\omega \sin \text{latitude} = f$ (usually), is known as the Coriolis parameter. Equation 3.25 becomes modified only by putting $\mathbf{Z} = \operatorname{curl} \mathbf{v} + f\,\mathbf{k}$. For motion whose horizontal scale is not large, $v_z \simeq v_{hor}$, but the Coriolis terms are negligible anyway.

3.15 Some other forms of the vorticity equation

Equation 3.25 can be written using the continuity of mass to eliminate $\operatorname{div} \mathbf{v}$

$$\frac{D\mathbf{Z}}{Dt} + \mathbf{Z}\operatorname{div}\mathbf{v} = \frac{D\mathbf{Z}}{Dt} - \frac{\mathbf{Z}}{\rho}\frac{D\rho}{Dt} = \rho\frac{D}{Dt}\left(\frac{\mathbf{Z}}{\rho}\right) \tag{3.31}$$

In the circulation theorem, we partially eliminated ∇p by a line integral of $\mathbf{v}.d\mathbf{r}$, but we would have completely eliminated the pressure term by making a line integral of $\rho\mathbf{v}.d\mathbf{r}$, which would lead us to expect conservation of something like $\operatorname{curl}\rho\mathbf{v}$. Equation 3.31 shows that it is something more like $(\operatorname{curl}\mathbf{v})/\rho$ which is conserved, with the variation with density in the inverse sense. Putting $\mathbf{N} = \mathbf{Z}/\rho$ equation 3.25 can be written

$$\frac{D\mathbf{N}}{Dt} - (\mathbf{N}.\nabla)\mathbf{v} + \frac{g}{\rho}\mathbf{k}\wedge\nabla\phi = 0 \tag{3.32}$$

Now, for some arbitrary function $q(x, y, z, t)$, consider the component of vorticity normal to the q-surfaces. Taking the scalar product of equation 3.32 with ∇q gives

$$\frac{D}{Dt}(\mathbf{N}.\nabla q) - \mathbf{N}.\frac{D}{Dt}\nabla q - \nabla q.(\mathbf{N}.\nabla)\mathbf{v} + \nabla q.\frac{1}{\rho}\nabla\rho\wedge\frac{1}{\rho}\nabla p = 0 \tag{3.33}$$

Somewhat surprisingly, the second and third terms can be combined and simplified. Notice that, in them, $\mathbf{N}$ is not differentiated but serves only to define a direction in space, $\mathbf{n}$ say. Therefore

$$\mathbf{N}.\frac{D}{Dt}\nabla q + \nabla q.(\mathbf{N}.\nabla)\mathbf{v} = \mathbf{N}\frac{\partial^2 q}{\partial t\partial n} + \mathbf{N}(\mathbf{v}.\nabla)\frac{\partial q}{\partial n} + \nabla q.\frac{\partial\mathbf{v}}{\partial n}$$

$$= \mathbf{N}\frac{\partial^2 q}{\partial t\partial n} + \mathbf{N}\frac{\partial}{\partial n}(\mathbf{v}.\nabla q) = \mathbf{N}\frac{\partial}{\partial n}(\frac{Dq}{Dt})$$

$$= \mathbf{N}.\nabla\frac{Dq}{Dt}$$

whence equation 3.33 becomes

$$\frac{D}{Dt}(\mathbf{N}.\nabla q) - \mathbf{N}.\nabla\frac{Dq}{Dt} + \nabla q.\frac{1}{\rho}\nabla\rho\wedge\frac{1}{\rho}\nabla p = 0 \tag{3.34}$$

This reproduces the vertical component of the vorticity equation, when we put $q = z$. With $q = p$, the third term vanishes, and the resulting equation shows that $\mathbf{N}.\nabla p$ can, in an inviscid system, be generated only by $\mathbf{N}.\nabla\omega$, where $\omega = Dp/Dt$ is the analogue to vertical velocity in pressure coordinates. The baroclinic term also vanishes for $q = \phi$, the log-potential-temperature, so the quantity $(\operatorname{curl}\mathbf{v} + 2\omega).\nabla\phi/\rho$, known as the potential vorticity, satisfies

$$\frac{D}{Dt}\left(\frac{(\text{curl}\mathbf{v} + 2\omega).\nabla\phi}{\rho}\right) = \frac{(\text{curl}\mathbf{v} + 2\omega)}{\rho}.\nabla\frac{D\phi}{Dt} \qquad (3.35)$$

when viscosity is negligible. By way of physical interpretation, we notice that the stretching term remaining in equation 3.32 acts on the field of potential temperature in the same way as on the vortex, to give the compact form of equation 3.35. The identification of a second conservative property of inviscid adiabatic flow was an important intellectual achievement of Ertel and Rossby. Eady noticed conservation of quasi-geostrophic potential vorticity in his 1947 paper, but when I noticed and was delighted by it, he dismissed it as an incidental result.

For flow of large scale, the main contribution to the potential vorticity arises from the vertical component of planetary vorticity multiplied by the variation in stratification, and the mean stratification multiplying the relative vorticity. Since these are also the most discriminating features of large-scale flow, it was not surprising to me that the relation might also be very useful.

For flow which is confined to one vertical plane, the relative vorticity is perpendicular to $\nabla\phi$, curl$\mathbf{v}.\nabla\phi$, vanishes identically, and the potential vorticity tells us nothing about the flow.

Chapter 4

Nearly horizontal atmosphere

4.1 Nature of approximation

to thine own self be true

The smallness of a term in an equation is a hint that the term might be omitted, but we must be careful because a quantity that is negligible in one equation, may not be in another, perhaps because it appears there multiplied by a large factor. This is because it represents a different process in the new context. Micawber commented that even a small excess of expenditure over income eventually led to disaster.

Mathematicians like to expand in powers of a small number ϵ say, and keep only the lowest orders. This often presents difficulties in physical interpretation because the terms in ϵ^2 often represent very different physics to those in ϵ. For example a wave of small amplitude ϵ advects wave properties at a rate proportional to ϵ^2, and often it is the transfer we are really interested in, not just the existence of the wave. Thus we find the shape of the motion pattern, represented by the correlation between different properties, interesting, as well as the amplitude.

The longer-term evolution of the amplifying waves found in unstable linearised systems, depends on the non-linear terms, because it is only these that prevent the unlimited increase in amplitude. In contrast, the amplitude of forced waves is often a more incidental property. It might be argued that this is untrue of breaking waves where non-linearity is an essential ingredient. However, even there, the details of the energy dissipation may be much less important than the fact of its existence. Ocean waves propagate energy towards the shore almost independent of the nature of the dissipation there. Where there is very little dissipation, reflection may become significant, so we would like to know that dissipation is, or is not, going to take place but that might depend little on the details. Indeed, because dissipation is often accompanied by a cascade to small scales, the question of how much dissipation might direct our attention to the more detailed study of a space scale that we had previously ignored. Ocean waves approaching a rocky shoreline makes a nice example.

We would like our simplified equations to have some integral conservation properties, similar to those of the real system. It would be comforting if the model conserved mass, and had an integral looking like energy conservation, with kinetic energy, potential energy, and similar properties represented in it. One way of achieving this sort of quality is to develop a set of equations that consistently describes a possible set of physical properties. These may not be realistic but must be consistent. For example, many aspects of tropospheric motion can be represented in a model in which the air is supposed to be of almost constant density as it moves to different pressures, in spite of the large variation of density with pressure in the real system. The incompressible system has the same number of integral constraints as the real system, even though quantitatively different.

4.2 Nearly horizontal atmosphere analysed

The atmosphere is a shallow layer of compressible, stratified gas, rotating very nearly with the solid Earth beneath. We are aware of sound waves, and gravity waves, and weather systems. Let us look for these in the fundamental equations. While doing so it is convenient to leave out various terms, and we will label these with tracer parameters whose correct value is unity, but which can be made to vanish one by one in order to gauge the nicety of a proposed omission.

It is convenient to work with p and $\phi = \log \theta$, rather than say, p and ρ, or p and T. We define a mean static equilibrium atmosphere $p_0(z)$, $\rho_0(z)$, $\phi_0(z)$, and suppose that the motion is associated with comparatively small variation about these mean values. Thus we put $p = p_0 + \delta p(x, y, zt)$ where $|\delta p/p_0| \ll 1$, $\phi = \phi_0 + \delta \phi$, where $|\delta \phi| \ll 1$, etc. and a little algebraic manipulation yields the following set

$$\frac{Dv_h}{Dt} + f k \wedge v + \nabla_h \frac{\delta p}{\rho} = \nu \nabla^2 v_h \tag{4.1}$$

$$n_4 \frac{Dw}{Dt} + \frac{\partial}{\partial z}\left(\frac{\delta p}{\rho_0}\right) - n_3 B \frac{\delta p}{\rho_0} = g \, \delta\phi + \nu \nabla^2 w \tag{4.2}$$

$$\operatorname{div} v - n_1 \frac{w}{H_0} + n_2 \frac{D}{Dt}\left(\frac{\delta\rho}{\rho_0}\right) = 0 \tag{4.3}$$

$$\frac{D}{Dt}\delta\phi + Bw = \kappa\nabla^2\phi \tag{4.4}$$

where $\delta\phi = \delta p/\gamma p_0 - \delta\rho/\rho_0$ and $B = d\phi_0/dz$ represents the static stability, $H_0 = -\rho_0/(d\rho_0/dz)$ is the density scale height, and v and κ are the kinematic molecular diffusivities for momentum and heat, respectively. Of the tracer parameters

 ▨ n_1 follows the mass-continuity effect
 ▨ n_2 follows the elastic compressibility
 ▨ n_3 is an adjustment parameter
 ▨ n_4 follows the non-static contribution to the pressure.

We notice that $1/H_0 = B + g/\gamma RT_0$. It is convenient to put $c^2 = \gamma RT_0$.

4.3 The linearised equations

There are enormous advantages in treating linear equations. They are relatively easy to solve, and their solutions can be generalised by being added together. This is true here where we consider perturbations of small amplitude first. This is restrictive because almost nothing happens if the amplitude is truly small. However, as Eady pointed out when being criticised for having produced only linear solutions, the dynamics of non-linear systems includes that of linear systems; the linear mechanics does not become invalid, but may be substantially modified in the non-linear regime. Finally, we might note that, in order to discover the contents of a mystery parcel, we first pick it up and shake it gently; we do not instantly embark on a violent non-linear dissection, just in case it contains something delicate that might get broken. In this section we suppose that all quadratic terms can be neglected relative to linear ones, thus neglecting almost all aspects of advection. Since meteorology is primarily concerned with such transport, this means that our set may not be directly applicable to real meteorological systems. Nevertheless it will give us some clues as to how these behave. Let us see what is left.

For simplicity we assume that the motion is two-dimensional, in the sense of being dependent on only two space dimensions (x, z). Consistently we neglect the variation of Coriolis parameter with latitude. We also ignore molecular diffusivity of both heat and momentum. The linearised equations, including the four tracer parameters n_1 to n_4 are

$$\frac{\partial v}{\partial t} + fu = 0 \qquad \frac{\partial u}{\partial t} - fv + \frac{\partial}{\partial x}\left(\frac{\delta p}{\rho_0}\right) = 0 \tag{4.5}$$

$$n_4 \frac{\partial w}{\partial t} + \frac{\partial}{\partial z}\left(\frac{\delta p}{\rho_0}\right) - n_3 B \frac{\delta p}{\rho_0} - g\delta\phi = 0 \tag{4.6}$$

$$\frac{\partial u}{\partial x} + \frac{\partial w}{\partial z} - \frac{n_1}{H_0} w + n_2 \frac{\partial}{\partial t}\left(\frac{\delta p}{\rho_0}\right) = 0 \tag{4.7}$$

$$\frac{\partial}{\partial t}\delta\phi + Bw = 0 \tag{4.8}$$

Since the coefficients are real and independent of x and t, sinusoidal solutions can be written in exponential form, with each variable proportional to $\exp\, i(kx + \sigma t)$. Thus we can replace $\partial/\partial x = ik$, $\partial/\partial t = i\sigma$ and eliminating in turn p, v, u, $\delta\phi$, gives

$$\left(\frac{\partial}{\partial z} + Bn_2 - \frac{n_1}{H_0}\right)w + \frac{i\sigma n_2}{c^2}\frac{\delta p}{\rho_0} - \frac{ik^2\sigma}{\sigma^2 - f^2}\frac{\delta p}{\rho_0} = 0 \tag{4.9}$$

$$(gB - \sigma^2 n_4)w + i\sigma\left(\frac{\partial}{\partial z} - Bn_3\right)\frac{\delta p}{\rho_0} = 0 \tag{4.10}$$

Finally, eliminating $\delta p/\rho_0$ and assuming that B, c, and H_0 can be replaced by suitable mean values independent of z, we get

$$\sigma\left\{\left(\frac{\partial}{\partial z} - Bn_3\right)\left(\frac{\partial}{\partial z} + Bn_2 - \frac{n_1}{H_0}\right) + (gB - \sigma^2 n_4)\left(\frac{k^2}{\sigma^2 - f^2} - \frac{n_2}{c^2}\right)\right\} = 0 \tag{4.11}$$

If no terms are neglected, we have $n_1 = n_2 = n_3 = n_4 = 1$. We notice that equation 4.11 is then of the fifth order in σ. Figure 4.1 shows the phase speed as a function of wavelength for values of parameters characteristic of the troposphere.

Waves of compression/rarefaction, as in sound waves, move quickly and may be of high frequency but the mechanism exists independent of whether or not they can be detected by the human ear. Gravity waves, as in the undulations of the free surface of a liquid, move slower, and very long waves have such a long period that they become dominated by the inertial period of the Coriolis acceleration. The longest gravity wave which also has a very long vertical wavelength moves as fast as the speed of sound, but is nevertheless a gravity wave. Internal gravity waves, like those that exist on the interface between two immiscible liquids of different density, are the ones with shorter vertical wavelengths. Waves with very long horizontal wavelength are dominated by the inertial effects of Coriolis again, but fairly long ones are dominated by the internal stratification much like long external waves, and have propagation speed independent of wavelength, for the same reasons. Short internal gravity waves have

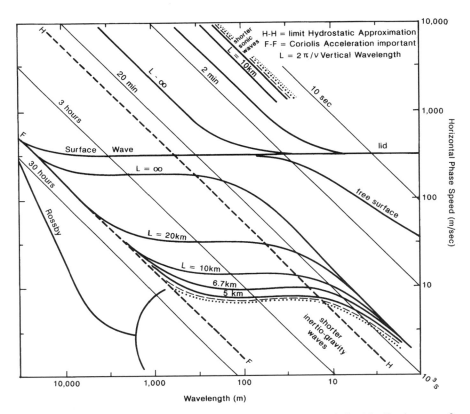

Figure 4.1 Variation of phase speed with wavelength for idealised types of waves using parameters typical of the troposphere. Waves of compression (sound waves) are fast, whether the human ear can hear them or not. Very long, or slowly oscillating, gravity waves are dominated by the Earth's rotation. The longest gravity wave with no phase change in the vertical moves as fast as a sonic wave, but we argue that it is a gravity wave. Waves with shorter vertical wavelengths are dominated by the internal stratification but the bulk of their energy is in the horizontal component of the velocitiy, leading to a propagation speed independent of wavelength. The response of the short waves is Archimedian and the frequency becomes independant of wavelength. Motion on the scale of days (synoptic scale) is nearly geostrophic but with timescale determined by other factors. Transport of fluid properties by waves is where the waves move at the same speed as the fluid. Thus there will be a band of interesting motion systems where the phase speed is comparable with a typical tropospheric wind speed of $10 \, \mathrm{m \, s^{-1}}$.

little pressure field, and are therefore Archimedian in character, with frequency independent of horizontal wavelength.

Motion of very large, or synoptic, scale is nearly geostrophic, but the propagation speed is determined by other factors.

Transport of properties over considerable distances can be achieved only by waves that move at velocity comparable with that of the fluid. Thus there are regimes of such advective systems where the phase speed is comparable with $10\,\mathrm{m\,s^{-1}}$. We now examine the mechanics of different parts of this picture.

4.3.1 Quasi-geostrophic wave

Referring back to equations 4.5 to 4.8, we see that the root $\sigma = 0$ gives purely horizontal, geostrophically balanced motion with hydrostatic pressure. The motion is non-dispersive in the sense that it does not depend on either the horizontal or vertical scale.

We identify this mode with the long-lived quasi-geostrophic motion observed on a space scale of 1000 km or so, equivilant to a wavelength of 6000 km, and which move with speeds typical of the mean flow. That the component waves are non-dispersive is consistent with the observation that the patterns, which we can express as wave packets, persist for a long time.

The remaining roots can be expressed nicely by supposing that the variables u, v, w, $\delta\phi$, and $\delta p/\rho_0$ depend on height like $\exp(i\nu + 1/2\,H_0)z$. We notice that this multiplying factor is essentially $\rho_0^{1/2}$, as it was in Figure 1.1, so the boundedness of the function $\exp i\nu z$ is practically the same as boundedness of the kinetic energy per unit volume; $\frac{1}{2}\rho v^2$, which makes good sense, and is a useful way of looking at observations of propagation over large ranges of height. Notice that we are now using ν as the vertical wavenumber, rather than a molecular coefficient of viscosity. This notation is a relic of Eady's excellent convention of using (λ, μ, ν) as a Cartesian triad of wavenumber. Unfortunately this was displaced by one-dimensionalists who promoted k as the only possible wavenumber, reserving λ for traditional use as wavelength, and leaving (k, l, m) for the only useful Cartesian triad. This substitution gives, whatever the notation, the remaining two roots

$$k^2\frac{(gB - \sigma^2)}{(\sigma^2 - f^2)} + \frac{\sigma^2}{c^2} = \nu^2 + \frac{1}{4H_0^2} \tag{4.12}$$

This set of equations is complete in the sense that it can be used to forecast from any initial state, defined as having u, v, w, p, and ϕ given at $t = 0$. We need only Fourier analyse the initial values weighted by $\exp z/2H_0$. The question of suitable spatial boundary conditions is not so clear, and will merit attention later.

4.3.2 Gravity inertial wave

For k sufficiently large, one root of equation 4.12 is $\sigma^2 = gB + O(k^{-2})$. Assuming realistically, that f^2/gB is small (typically $\sim 10^{-4}$ in the troposphere) the thermodynamic equation gives $\delta\phi \simeq (iB/\sigma)w$, and the vertical component of momentum, equation 4.6 expresses balance between the vertical acceleration $\partial w/\partial t$ and buoyancy $g\,\delta\phi$, with pressure negligible relative to buoyancy with ratio proportional to $(kH)^{-2}$. This is the Archimedian solution associated with the names of Brunt and Vaisala. It is sometimes convenient to recognise it with the special symbol N, where $N^2 = gB$. We notice that the static stability B, appears in two distinct roles; as a component of N the buoyancy frequency, and as a part of the variation of density with height as it appears in the momentum equation. It is usually clear from the context which one of these physical properties is being specified. For k sufficiently small, one root of equation 4.12 is $\sigma^2 = f^2 + O(k^2)$. The motion is purely horizontal and neither pressure nor buoyancy forces are important. Brunt calls this motion inertial. While something of a meteorological curiosity, some mesoscale motion may be represented. In the early evening, when there is not a strong wind, the surface friction is cut off from the air above by the developing nocturnal boundary layer. This force is thereby removed and the air responds through oscillations of inertial frequency. This is evident on some occasions as a 'nocturnal jet'. A reveiwer quotes inertial waves as common in the lower stratosphere; I wonder how they are detected with coarse temporal resolution. The relatively large phase speed of long inertial waves is likely to be computationally, even if not dynamically, significant. It is not glaringly obvious to me that these two extremes of wavelength are for the same wave, but they are, and define the wavelength limits of the gravity-inertial wave shown in Figure 4.1 .

4.3.3 Sonic wave

For c^2 large there is a root

$$\sigma^2 = (k^2 + v^2 + 1/4H_0^2)c^2 \tag{4.13}$$

This wave propagates, nearly at the speed of sound, in the direction normal to the lines of constant phase. It is nearly a longitudinal wave with particle velocities mainly along the direction of propagation, and perturbations of the entropy (due to the Bw-term) are comparatively small.

4.4 Boundary conditions

The vertical extent of the motion is limited by the nature of the upper and lower boundaries. These boundaries are often arbitrary, in that the true system is really unbounded, and we limit the extent of our study by trying to confine our attention to a limited part of space. However there are some interesting limit cases.

4.4.1 Rigid-lid boundary

Demanding that there is a lid, at a finite height, at which the vertical component of velocity vanishes, quantises possible values of the vertical wavenumber ν. However equations 4.9 and 4.10 show two other possibilities. We can have $\sigma = 0$, and $w = 0$ everywhere, which is exactly geostophic motion. Or we can have $\sigma^2 = c^2 k^2 + f^2$ and $w = 0$ everywhere, which is more interesting. It represents a sound wave, modified by rotation, propagating horizontally between the two lids. The rigid lid supports a pressure force, which varies with height like $\exp Bz$, which, for meteorological values, means rather little. Thus we see a pressure pulse propagating down a channel defined by the lid and the ground pressing on these bounding surfaces as it goes.

4.4.2 Free surface boundary

An alternative condition supposes that there is no more fluid above some level; this free surface may move in response to the motion, but cannot support a pressure. In this sense it is an opposite condition to the rigid lid. The condition is expressed mathematically by $Dp/Dt = 0$ at the upper boundary, which is represented by a surface $z = h(x, y, t)$, which moves with the fluid. This therefore satisfies $w = Dh/Dt$ at $z = h$. The linearised pressure relation gives $i\sigma \delta p = \rho_0 g w$ which combined with equation 4.9 gives $\partial w/\partial z = k^2 g w/(\sigma^2 - f^2)$ at $z = h$ as the condition to be satisfied by w. A general solution is messy, but no essential features are lost by assuming the pressure hydrostatic, which is good anyway if the waves are long enough, and consistently neglecting the effect of static stability by putting $B = 0$. We find that w is then proportional to $1 - \exp z/H_0$ and $\sigma^2 = f^2 + k^2 g H_0(1 - \exp -h/H_0)$. This solution reduces to that for long waves on the free surface of a liquid for $h \ll H_0$, and also shows that such surface waves continue to exist even for $h \gg H_0$, when the density at the free surface is vanishingly small.

4.5 Application to the real atmosphere

Later on, we will study some of these wave-types in more detail, by making approximations that are more appropriate to their individual needs. For the moment let us note that gravity-inertial waves and sound waves, are properties of the deep atmosphere, slightly quantised by the spherical geometry, and by boundary conditions in the vertical. Do we see evidence of these in the real atmosphere? Loud bangs at the ground, and smaller ones from supersonic aircraft, propagate anomalously great distances when they are reflected off the region of the upper stratosphere where the temperature increases with height, thereby reflecting sonic waves. Night-time conversations, especially near water, are readily ducted between a nocturnal inversion and a smooth water surface.

Some layers of the atmosphere are unable to transmit, and therefore must reflect, gravity waves. Astonishing at first sight, these are layers where the stratification is less stable. These layers behave as lids trapping the gravity waves, and are sometimes evident on satellite pictures showing extensive trains of gravity waves trapped in the lee of mountains.

Sometimes energy propagates to great heights where a number of things may happen to it. Equation 4.12 predicts that the pressure driving the wave motion should decrease with height like $\rho^{1/2}$, whereas the ambient pressure decreases more rapidly, like ρ. Thus the wave might run out of ambient pressure. The governing physics might well be represented by a free-surface boundary condition at a level somewhere near where the ambient and wave pressures became comparable. We suspect that this would depend on the vertical wavelength compared with the density scale height.

Alternatively, large vertical displacements may lead to adiabatically generated temperature contrasts that could be diminished through electromagnetic radiation in the terrestrial waveband, thereby damping the wave. The kinematic viscosity μ/ρ increases with height and one would expect it to dissipate motion of given temporal frequency. Curiously the opposite may happen, and viscosity *aid* the propagation by making succesive layers of air acquire the same horizontal velocity, just like a waiter carrying a pile of plates. But if the fluid speeds become independent of height, instead of increasing like $\rho^{-1/2}$ then this is damping.

Molecular conductivity of heat has a similarly unexpected effect. In an upward propagating sound wave, temperature contrasts arise through the adiabatic compressions and rarefactions of the air. Thermal conduction tends to erase these contrasts and thereby to dampen the wave motion. However, this mechanism finally converts the wave into a wave with isothermal expansions, not now vulnerable to dissipation by conduction, and which propagates onward at the isothermal, or Newtonian, speed of sound. Some energy is lost during the transition, but the motion is lossless thereafter.

The wave becomes more susceptible to shearing instability as the fluid velocities increase and the isentropes become more inclined to the horizontal. It might become vulnerable to irreversible breaking as the amplitude increases.

Some gravity waves, generated by the diurnal variation of solar heating in the stratosphere, reach heights of about 120 km where they begin to pose problems like this.

Sometimes one boundary condition may be easier to handle computationally than the other and be adopted for this less satisfactory reason. For example a free surface is natural in pressure coordinates, but a rigid lid is more natural in height coordinates.

Perhaps the least objectionable boundary condition to put at the upper limit of a model is that of no returning energy (at its simplest, the radiation condition) but this is rather difficult to formulate and to apply, especially in a highly dispersive medium where the group velocity is much different to the phase velocity. Like with the weather forecast models of limited area, one has the impression that, if a boundary condition is important then we ought to be modelling what happens on the other side of it anyway.

4.6 What do we expect to see?

Meteorology is largely concerned with advection; the transport of heat and other properties over large distances. In order to carry air long distances a wave must be of large amplitude, but it must also move with the fluid which is being advected. We therefore expect to find that it is the waves that move at speeds comparable with that of the air that are important in the real atmosphere. In theoretical discussions this is illustrated by the so-called critical level instability, which is the level in the fluid where the real part of the phase velocity coincides with the fluid velocity, and preferably, the growth rate is small. Taking air-speeds of about $10\,\mathrm{m\,s^{-1}}$ as typical, Figure 4.1 suggests that we should find quasi-geostrophic motion of $10\,000\,\mathrm{km}$ wavelength, inertial waves with vertical wavelength of a few kilometres and horizontal wavelength of 10 to $1000\,\mathrm{km}$, and gravity waves of horizontal wavelength $10\,\mathrm{km}$.

4.7 Simplified solutions

We want to simplify the equations so as to make them easier to handle, and therefore capable of tackling more realistic problems. We would like for example, to relax the poor assumption of negligible advection.

We would like our simplified set to reproduce the wave speeds of the full set, as well as having a 'nice look' to them – which implies something about conservation properties. Solving equations 4.5 to 4.8 as before, except keeping in the tracer parameters n_1 to n_4, we find that several terms contain the factor $(n_2 - n_3)$. This vanishes in the exact system with all $n_i = 1$, and will also vanish in the inexact set if we put n_2 to vanish at the same time as n_3. For brevity we therefore set $n_2 = n_3$ from the start. Quite why this happens is not clear to me, and I find it annoying to introduce such an arbitrary correction factor, but maybe if we thought about it more carefully we could convince ourselves it was a justifiable step. We also substitute, from the start, and much more justifiably

$$\frac{\delta p}{\rho_0} \propto \exp\left(iv + \frac{n_1}{2H_0}\right)z \qquad (4.14)$$

which gives

$$v^2 + \frac{n_1^2}{4H_0^2} + k^2\frac{(\sigma^2 n_4 - gB)}{\sigma^2 - f^2} - \frac{\sigma^2 n_2 n_4}{c^2} + Bn_2\left(\frac{1 - n_1}{H_0} + (n_3 - 1)B\right) = 0$$

$$(4.15)$$

This takes the place of equation 4.12, to which it reduces when all the tracers are set equal to unity.

4.8 Elastic waves eliminated

For many applications the evolution of the motion is so slow that the finiteness of the speed of sound is not important. To put $c \to \infty$ would eliminate sound waves from equation 4.15, but it would be an inaccurate and clumsy device. For example, as mentioned earlier, we might have some doubt about how to treat the relation $B + g/c^2 = 1/H_0$ which connects the different physical processes of sound propagation, inertial density, and stratification. It is more convenient to eliminate elastic waves as a physical process. Now sound waves represent a feedback loop in the equation-set where pressure gradient $\to$ acceleration $\to$ divergence $\to$ adiabatic change of pressure $\to$ pressure gradient. Breaking these connections anywhere breaks the feedback loop and therefore eliminates the elastic wave. For example, leaving out the term equating the horizontal acceleration of the air due to horizontal gradients of pressure, eliminates horizontally propagating sound waves. Unfortunately it also eliminates a great deal of meteorologically interesting motion at the same time.

However we notice that the term $D(\delta\rho/\rho)/Dt$ in the mass continuity equation 4.3, is numerically small for meteorological systems (about 1% of $\partial w/\partial z$) but is certainly part of the feedback loop supporting sound waves. This term can be omitted by setting n_2 to zero. Notice that this makes the last term in equation 4.15 vanish, which is a good thing because it would also vanish in the full set, where $n_1 = 1$, and $n_3 = 1$.

We expect the term in equation 4.15 containing c^2 to be least negligible at high frequencies, in which case the criterion for justifying its neglect is

$$k^2 + v^2 + n_1^2/4H_0^2 \gg \sigma^2/c^2 \qquad (4.16)$$

which is that the phase speed of the motion should be small compared with the speed of sound, after they have been raised to the second power. This is a good solid approximation for most atmospheric motion. Moreover it happens that this criterion is rather stiff and can be violated when the elastic process is still negligible. Perhaps the most important class of motion in this category is the daily atmospheric tide, which happens to move round the earth at about the speed of sound. This is undoubtably a long gravity wave for which the elastic mechanism is negligible. Consistently it is not a rapid oscillation in the sense we assumed above, in order to give a nice neat criterion. This near resonance of the diurnal tide may be a coincidence, but there may be connections with the momentum transfer of the gravitational tide and the speed of rotation of the planet on a very much longer time scale.

The only remaining root of equation 4.15 is

$$(n_4 k^2 + v^2 + n_1^2/4H_0^2)\sigma^2 = (v^2 + n_1^2/4H_0^2)f^2 + k^2 N^2 \qquad (4.17)$$

This agrees with the exact solution for k^2 large (buoyancy oscillations) and for k^2 small (inertial oscillations). For the range

$$v^2 f^2/N^2 \ll k^2 \ll v^2 \qquad (4.18)$$

the horizontal phase speed is nearly independent of the horizontal wavelength. Since $f/N \sim 10^{-2}$ this is usually a considerable interval. The solution in this range is the internal-wave analogy to the surface gravity wave which is long in comparison with the depth.

There is no solution for the rigid lid wave, which has $w = 0$ at all levels. This is consistent with this wave being essentially a sound wave, and therefore eliminated by setting $n_2 = 0$.

There is a solution for the free surface wave. For example, in the not-unrealistic situation $\sigma^2 \gg N^2 \quad \sigma^2 \gg f^2$, we find

$$\sigma^2 = gk \tanh kh, \quad w \propto \sinh kz \qquad (4.19)$$

just as for surface gravity waves on a free surface of depth h. There is something inherently unsatisfactory about these waves juggling the upper (fictitious) surface, just to provide the pressure field for the motion below. In some sense the lid condition is preferable.

4.9 Compressible-Boussinesq

Putting $n_2 = n_3 = 0$, in the full non-linear equations gives the compressible analogue to the Boussinesq approximation for a liquid. Density variations are neglected except

(i)   where multiplied by g,
(ii) possibly where differentiated w.r.t. height in mass continuity,
(iii) to convert the natural variable $\delta p/\rho_0$ into pressure perturbation.

We note that the thermodynamic equation is conveniently expressed in terms of potential temperature and not gas-kinetic temperature. We explore further integral and differential properties of this set in Chapter 5.

4.10 Hydrostatic approximation alone

Putting $n_4 = 0$ forces the pressure to be equal to the hydrostatic value. Since this causes the term in c^2 to disappear from equation 4.15 it follows that internal sound waves are eliminated. Boundary sound waves need special discussion which will follow later. Terms identified by the tracer parameter n_4 in equation 4.15 can certainly be neglected if $\sigma^2/c^2 \ll (k^2 + v^2 + 1/4H_0^2)$ and $k^2 \ll v^2$. The first is typically well satisfied and is the same criterion as for the incompressible approximation, but the second criterion is more restrictive. Rigid lid and free surface boundary solutions still exist, though the free surface waves are distorted at short wavelengths indicated by the second criterion above, by not varying exponentially with depth. The horizontally propagating sound wave found with rigid-lid boundaries remains with the hydrostatic approximation.

4.10.1 Hydrostatic pressure and no sound waves

Putting $n_4 = n_2 = n_3 = 0$, gives the same frequency equation 4.16, with $n_4 = 0$. The rigid-lid wave does not now appear, but the free surface condition gives a very spurious wave

$$\sigma^2 = f^2 + k^2 g H_0 (\exp h/H_0 - 1) \tag{4.20}$$

unless the inertial density is also eliminated.

4.10.2 Variation of mean density with height

The upward decrease of mean density, that distinguishes between continuity of mass and of volume, is represented by the tracer parameter n_1 and appears in the multiplier $\exp z/2H_0$ and the term $1/4H_0^2$. The multiplier is significantly greater than unity for depths comparable with that of the troposphere, whereas the term $1/4H_0^2$ appears added to v^2 so can certainly be neglected whenever the vertical wavelength is not much greater than the depth of the troposphere. What this means is that the variation of density with height can be neglected as long as the density factor ρ_0 is included implicitly. Thus the predicted value of u say using the set of equations in which the variation of mean density has been neglected, should be compared with $\rho_0^{1/2} u$ observed in the real atmosphere.

4.11 Quasi-geostrophic motion and Rossby waves

The motion corresponding to the root $\sigma = 0$ of equation 4.11 is purely advected and the mechanics is therefore independent of the values of the tracer parameters. More general solutions are not purely advected and are not independent of tracers. Suppose that such waves move much more slowly than sound waves, and that the hydrostatic approximation is acceptable, but that the gravity inertial mechanism might not be negligible. Thus we put $n_2 = n_3 = n_4 = 0$. As before, we assume the solutions separable in x and t, but now demand that y is directed from south to north and that the variation of the Coriolis parameter with latitude may be important. Solving equations 4.5 and 4.6 for u and v in terms of $\delta p / \rho_0$ gives

$$u = \frac{\sigma k}{f^2 - \sigma^2} \frac{\delta p}{\rho_0} \qquad v = \frac{\mathrm{i} f k}{f^2 - \sigma^2} \frac{\delta p}{\rho_0} \qquad (4.21)$$

while $\delta \phi$ eliminated between equations 4.7 and 4.9 gives

$$\frac{\partial}{\partial z}\left(\frac{\delta p}{\rho_0}\right) = \mathrm{i}\left(\frac{N^2}{\sigma}\right) w \qquad (4.22)$$

These velocities are substituted in the equation of continuity of mass, but the term in $\beta = \partial f / \partial y$ is now kept in to give

$$\left(\sigma \left(k^2 + \frac{n_1}{H_0} \frac{(f^2 - \sigma^2)}{N^2} - \frac{(f^2 - \sigma^2)}{N^2} \frac{\partial^2}{\partial z^2} \right) - k\beta \frac{(f^2 + \sigma^2)}{(f^2 - \sigma^2)} \right) \frac{\delta p}{\rho_0} = 0 \qquad (4.23)$$

and we have assumed that suitable mean values of f^2 have been substituted for $f^2(y)$. This simplification can be relaxed and gives ther possibility of waves trapped near the equator, instead of waves trapped by bogus boundaries at the imitation poles.

Equation 4.23 is of the fifth order in σ, and since we have already eliminated sound waves we have acquired two more solutions. This is not good physics. However, the oscillations in systems of large scale are usually so slow as to suggest that $\sigma^2 \ll f^2$, which discriminates against the extra pair of solutions. The form of the resulting equation again suggests the substitution

$$\delta p / \rho_0 \propto \exp(\mathrm{i}\nu + n_1/2H_0)z \qquad (4.24)$$

whence

$$\sigma = \frac{-k\beta}{k^2 + (f^2/N^2)(\nu^2 + n_1^2/4H_0^2)} \qquad (4.25)$$

The density-scale-height appears exactly as in the sound- and gravity- wave solutions, and as there, is negligible in the frequency equation, apart from waves with extraordinarily long vertical and horizontal wavelengths.

Equation 4.25 is the usual form for Rossby waves. We have $\sigma \simeq -\beta/k$ so the criterion $\sigma^2 \ll f^2$ is satisfied if $(\beta/fk)^2 \ll 1$ corresponding to wavelengths $(2\pi/k)$ shorter than some 20 000 km. But by definition $\beta \simeq f/L$, where L is comparable with the radius of the Earth so this result becomes-that the planetary wave of largest possible wavelength is on the verge of being non-geostrophic. More accurate analysis using spherical geometry confirms this curious result.

When variation in the W–E current U is included, the timescale of particles passing through the wave includes a contribution like $k(\Delta U - c)$ due to advection of the fluid relative to the wave. The criterion $\sigma^2 \ll f^2$ then also gives a short wavelength limit to geostrophy. But the value 2000 km given by a typical relative speed of $10\,\mathrm{m\,s^{-1}}$ may make this lower limit too large, because shorter waves are likely to penetrate less far in the vertical than longer, so detect less velocity difference than longer waves.

Equation 4.23 has a root close to $\sigma^2 = f^2 - 2\beta f/k$ for $\beta \to 0$ which may be spurious. The root $\sigma^2 = f^2 + N^2 k^2/v^2$ is not spurious, but shows that long inertial-gravity waves are not excluded from the system. This we might view as a defect if, as seems likely, these waves have little effect on the nearly geostrophic systems on which we will now focus attention.

Let us explore how this class of waves might be avoided. If we assume $\sigma^2 \ll f^2$ in equation 4.23 they are eliminated, but at the expense of leaving out the term in k^2 in equation 4.25, which is not good. We notice that the criterion implies that the horizontal wind is so accurately geostrophic that it may be used to estimate the horizontal divergence of the real wind. This is not a good assumption for the atmosphere. It is better for the ocean where it forms the main step in describing the Sverdrup flow which is enormously slow, and is discussed in Section 8.4.1.

If we use the geostrophic wind in the advection terms, but not to find the horizontal divergence, a little manipulation shows that we reproduce the solutions of equation 4.23 but without the gravity waves. Generalisation of this procedure leads us to the quasi-geostrophic set, which is basic to many theoretical analyses of large-scale motion. If we retain the true divergent winds in the advection terms we finish up with the semi-geostrophic set.

4.11.1 Two-dimensional motion; Kelvin waves

So far we have considered only motion dependent on only one horizontal dimension. Waves in two horizontal dimensions generally have k^2 replaced by $k^2 + \mu^2$ where μ is the y-wavenumber. This is a fairly trivial modification, unless μ is purely imaginary, when the total wavenumber $k^2 + \mu^2$ is reduced. In this situation the stream function, being exponential in form, is unbounded on one side, so must be limited in some way. The classic case concerns surface gravity waves propagating parallel to a rigid boundary. We easily see that, in the presence of the Coriolis effect, waves in which the y-component of velocity vanishes everywhere are possible, so long as the pressure gradient varies as $\exp(-fy/c)$ where $c = (gH)^{1/2}$ is the speed of the gravity wave. Thus, in the northern hemisphere, these waves can propagate along the boundary to its right so long as c is positive. If c is negative then the wave can propagate only along the boundary to its left. Are these distorted gravity waves or something fundamentally different? What happens to the wave trying to propagate in the forbidden direction?

We are dealing essentially with distorted gravity waves so imagine a region of excess mass suddenly created near the wall. At first it will behave as a gravity wave spreading into the fluid further away, going equally in both directions along the boundary. After a time of order f^{-1} Coriolis accelerations becomes tangible and we suppose that the wave propagating to the right will tend to drift towards the boundary, becoming more like the exponential variation of the Kelvin wave. The wave propagating left might drift away from the boundary, with maximum amplitude not at the boundary. Perhaps the amplitude between the boundary and the maximum increases with distance like $\exp(+fy/c)$. Apart from complications due to the variation of Coriolis parameter with latitude, the analysis is fairly straightforward. Linearised free surface equations are

$$\frac{\partial u}{\partial t} - fv + g\frac{\partial h}{\partial x} = 0$$

$$\frac{\partial v}{\partial t} + fu + g\frac{\partial h}{\partial y} = 0$$

$$H\left(\frac{\partial u}{\partial x} + \frac{\partial v}{\partial y}\right) + \frac{\partial h}{\partial t} = 0 \tag{4.26}$$

Solutions proportional to $\exp i(kx + \mu y + \omega t)$ are possible so long as

$$\omega^2 = (k^2 + \mu^2)gH + f^2 \tag{4.27}$$

and satisfy the condition v = 0 at y = 0 so long as

$$h = \exp i(kx + \omega t)\left(\cos \mu y + \frac{kf}{\mu\omega}\sin \mu y\right) \tag{4.28}$$

while the other solution has the sign of ω reversed. For large time, there is a point of stationary phase at $\mu = 0$, and expanding about there gives

$$\omega = \omega_0 + (gH/2\omega_0)\mu^2 + O\mu^4, \text{ where } \omega_0^2 = f^2 + k^2gH \tag{4.29}$$

For ω positive the path of steepest descents is in the direction $(1 + i)$ whereas for ω negative it is $(1 - i)$ giving contributions to the Fourier integral like $t^{-1/2}e^{i\omega t}(1 + kfy/\omega_0)$ and $t^{-1/2}e^{-i\omega t}(1 - kfy/\omega_0)$, respectively, where terms in $\omega^2 y^2$ have been neglected. The first solution represents the one that *increases* exponentially with y, the second the one that *decreases* exponentially with y, at least for distances small compared with ω_0/kf. Outside the range of convergence, i.e. for $y^2 \sim \mu^2 \sim gHt/2\omega_0$, the wave energy has not yet arrived, so the solution must be of small amplitude.

Thus near $y = 0$ the solution behaves rather like a standing wave, the positive exponential part propagates into the fluid, and the negative exponential part remains near the boundary. While this begins to give the nature of the solution, it does not represent the physical problem we envisaged, because equation 4.28 does not have vanishing velocity at $t = 0$. What this means is easiest to appreciate in the one-dimensional case in which k is vanishingly small. Unfortunately this begs another question because we then cannot distinguish which way the wave propagates any longer. Never mind, it gives some additional information.

The symmetrical solution is the sum of equation 4.28 for positive and negative values of ω, $h = \cos \mu y \, \cos \omega t$, $u = (Af/\mu H) \sin \mu y \cos \omega t$. This does not satisfy the condition $u = 0$ at $t = 0$, but we now can see that adding a component $\Delta h = \cos \mu y$ adds a term in $\sin \mu y$ to u and nothing to v, so can be used to give the solution which is initially stationary. This is

$$h = A(\cos \omega t + (f^2/\mu^2 gH)) \cos \mu y \tag{4.30}$$

and the Fourier synthesis for the complete solution is

$$h = \int A(\mu) \frac{(\cos \omega t + f^2/\mu^2 gH)}{(1 + f^2/\mu^2 gH)} \cos \mu y \, d\mu \tag{4.31}$$

where the singularity represents the Kelvin wave.

Figure 4.2 shows a computed example. At $t = 0$ a Gaussian hump of water of width 2, is initially centred on $y = 4$ away from the wall. Initially the gravity wave mechanism, with nomalised phase speed unity, dominates, and the hump splits into two parts going in opposite directions. Thus, by $t = 10$ we see two semi-humps, one near $y = 14$ which has travelled the 10 units outward from $y = 4$, and the other near $y = 6$, having travelled the 4 units to the wall then 6 units away from it after reflection. There is also a small residual trapped at the boundary, which is the Kelvin wave.

By $t = 20$ the trapped wave has a length scale similar to that of the Kelvin wave. The total mass of water is conserved. At $t = 0$ there was one peak of halfwidth $\Delta y = 4$ maximum amplitude 0.5. At $t = 20$, there are two peaks each of halfwidth $\delta y = 4$, amplitude 0.2 together with the part of amplitude 0.2 at the shore diminishing out to the 'Kelvin' scale $y = 10$, giving the mass sum; $4 \times 0.5 = 2 \times 4 \times 0.2 + 10 \times 0.2$ (nearly).

Figure 4.3 shows the same experiment but now in terms of the height of the free surface at the boundary wall, where we can see the arrival of the half-hump at $t = 4$, settling down to the Kelvin residual as time advances. One way of visualising the significance of this state is that it retains information near the boundary that would otherwise propagate away from it.

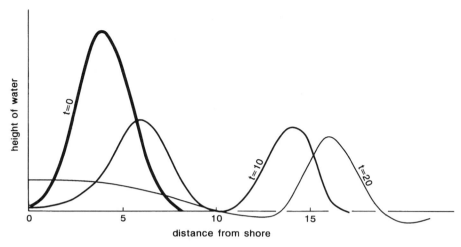

Figure 4.2 A Kelvin wave that results from the spreading out of a hump of water initially near a rigid boundary. Height of water is in arbitrary units. Distance scale is in units such that the halfwidth of the initial Gaussian hump is two, and the timescale such that the phase speed of gravity waves is unity. We see that by $t = 20$ the trapped wave has a length scale similar to the Kelvin wave.

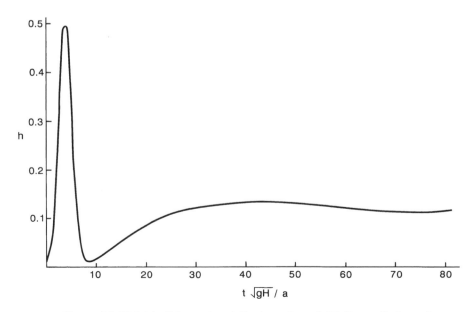

Figure 4.3 Height of the water at the boundary. Initially small, it reaches a peak at $t = 4$, as the half-lump hits it, then settles down to the residual value.

We can even visualise what is going on, for as the hump of water spreads out away from the wall, there is a net y-displacement that generates a velocity component parallel to the boundary, through $\partial u/\partial t = fv$ that will support a pressure gradient through the Coriolis force. To further the analysis one must distinguish between the displacements of the fluid δy and the scale of the variation Δy say. This way we argue that $\delta h \sim H\delta y/\Delta y$, $u \sim f\,\delta y$, $g\delta h/\Delta y \sim fu$, finally, $\Delta y \sim (gH)^{1/2}/f$ which, not surprisingly recovers the 'Rossby' adjustment scale for the attainment of geostrophic balance.

Chapter 5

Gravity waves

5.1 More realistic gravity waves

Here we neglect the direct effect of the Coriolis acceleration; assuming $\sigma^2 \gg f^2$, which implies a horizontal wavelength somewhat less than say 200 km. We also neglect the direct effect of molecular diffusion, which implies wavelengths more than about 10 cm. We neglect the variation of density with height where it appears as inertia by putting $n_1 = 0$, and concentrate attention on motion dependent on only two Cartesian coordinates. The first restriction is not important; the second is more serious. However, the difficulties involved in even just picturing motion in three dimensions are so great that we need the simpler concepts developed in two dimensions as a guide anyway. Our equation set is

$$\frac{D\mathbf{v}}{Dt} + \nabla\left(\frac{\delta p}{\rho_0}\right) - g\mathbf{k}\,\delta\phi = 0 \tag{5.1}$$

$$div\ \mathbf{v} = 0 \tag{5.2}$$

$$\frac{D}{Dt}(\delta\phi) + Bw = Q \tag{5.3}$$

where the variables depend only on (x, z, t). The y-component of equation 5.1 shows that $Dv/Dt = 0$; the component of velocity in the y-direction is conserved by individual parcels. Since v does not occur in any other equation it plays no further role in the analysis.

Taking the curl of equation 5.1 gives the vorticity equation

$$D\eta/Dt + g\ \partial\,\delta\phi/\partial x = 0 \tag{5.4}$$

where $\eta = \partial u/\partial z - \partial w/\partial x$ is the y-component of the vorticity. Equations 5.2, 5.3 and 5.4 are sufficient to define the system so long as the boundary conditions do not demand knowledge of the pressure. An elegant way of reducing the number of dependent variables is by defining a stream function ψ through $u = \partial\psi/\partial z$, $w = -\partial\psi/\partial x$ which can be thought of as the general solution to equation 5.2. We then get $\eta = \partial^2\psi/\partial x^2 + \partial^2\psi/\partial z^2$ which is nice and the set has a nice energy integral too. Thus taking the scalar product of equation 5.1 and velocity gives immediately

$$\frac{D}{Dt}\left(\frac{1}{2}\mathbf{v}^2\right) + \mathbf{v}.\nabla\left(\frac{\delta p}{\rho_0}\right) = gw\,\delta\phi \tag{5.5}$$

where the first term is related to the generation of kinetic energy, the second to work done by a pressure field, and the third to potential energy. This equation is concerned with mechanical energy. We notice, in contrast to the exact energy equation, that the small pressure disturbance δp replaces the full pressure in the exact but uncompromising energetics of Section 3.9.

Multiplying equation 3 by $\delta\phi$ gives

$$\frac{D}{Dt}\frac{1}{2}(\delta\phi)^2 + Bw\,\delta\phi = Q\,\delta\phi \tag{5.6}$$

which expresses conversion of thermal energy, represented by diabatic heating Q, into potential energy of a parcel represented by the first term, and potential energy being converted into mechanical energy in the second.

Eliminating this conversion $w\,\delta\phi$ between equations 5.5 and 5.6, gives an energy equation connecting kinetic and available potential energy

$$\frac{D}{Dt}\left(\frac{1}{2}\mathbf{v}^2 + \frac{g\,(\delta\phi)^2}{B}\right) + \mathbf{v}.\nabla\left(\frac{\delta p}{\rho_0}\right) = gQ\,\frac{\delta\phi}{B} \tag{5.7}$$

Another way of writing this comes from putting

$$w\,\delta\phi = D/Dt(z\,\delta\phi) - z\,D\phi/Dt \tag{5.8}$$

where we recognise that the potential energy available to this motion is $g\,z\,\delta\phi$, less by the small factor $\delta\phi$, than the full gravitational potential energy gz of Section 3.9.

5.2 Adiabatic perturbation of steady flow

We see that $u = U(z)$, $B = B(z)$ is an exact solution of the set of equations 5.1 to 5.3, and that it is not very interesting. The class of small perturbations about that state is more interesting. Putting $\psi = \psi_0 + \psi'$, $\delta\phi = \phi'$ and neglecting all product terms in primed quantities, we get

$$\left(\frac{\partial}{\partial t} + U\frac{\partial}{\partial x}\right)\nabla^2\psi' - \frac{d^2 U}{dz^2}\frac{\partial\psi'}{\partial x} + g\frac{\partial\phi'}{\partial x} = 0 \tag{5.9}$$

$$\left(\frac{\partial}{\partial t} + U\frac{\partial}{\partial x}\right)\phi' - B\frac{\partial\psi'}{\partial x} = 0 \tag{5.10}$$

When U and B are independent of z, these equations have simple wavelike solutions, and these define a complete set, in the sense that any initial value problem given ψ' and ϕ' at $t = 0$ can be solved in terms of a weighted sum of them. Thus $\psi' = a \exp i(k(x - ct)\pm vz)$ is a solution so long as

$$(U - c)^2(k^2 + v^2) = gB = N^2 \tag{5.11}$$

Already there is something interesting. If k and v are both real, the quantity $(k^2 + v^2)$ represents the wavenumber of the tilted wave $\exp i(kx + vz)$, and the phase speed depends only on this symmetrical combination in spite of the great dissimilarity in the mechanics between horizontal and vertical directions. If this wavenumber is sufficiently large then $c \simeq U$ which states that a wave which is short in either direction gets blown along by the wind.

5.3 Refraction and reflection

Suppose we attempt to propagate a wave of x-wavenumber k moving at speed c in one layer with $B = B_0$ say, into another layer with $B = B_1$ say. We see that, if $B_1 = B_+$ is larger than $B_0 = B_-$, as in Figure 5.1, v_1 will be real, and we might speak of the wave as being refracted. There might be a reflected component as well. But if B_1 is small the solutions in layer 1 may have v imaginary which

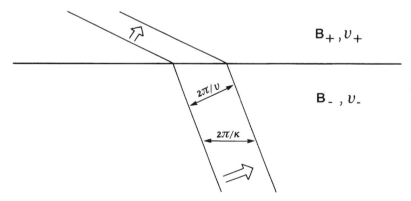

Figure 5.1 Simple oblique gravity wave passing from one medium to another. We note that if k is the x wavenumber and v is the z wavenumber, then $v = \sqrt{(k^2 + v^2)}$ is the wavenumber of the tilted wave. When the upper layer is more stably stratified than the lower then v is greater in the upper layer and the wave is refracted as shown.

implies that the wave amplitude must vary exponentially with distance into the second medium. The only acceptable solution is that the wave amplitude should *decrease* exponentially into the second medium, and that there should only be a reflected wave. We see that the second layer behaves like a perfect reflector of the wave. That the wave penetrates distance $1/|\nu_1|$ into the second layer suggests how deep this must be to approach perfect reflection.

If U is the same in the two layers, then the physical requirement of continuity of interface and pressure demands that ψ' and $\partial\psi'/\partial z$ must be continuous at the interface. When U is not the same, the interface is dynamically unstable, as discussed in Section 6.1. The condition that the phase speed must be the same in both layers gives

$$\nu_1^2 = \nu_0^2 + (N_1^2 - N_0^2)/(U-c)^2 \tag{5.12}$$

Putting

$$\psi' = \exp i\nu_0 z + R \exp -i\nu_0 z \text{ for } z < 0 \tag{5.13}$$

for the incident plus reflected wave, and

$$\psi' = T \exp i\nu_1 z \text{ for } z > 0 \tag{5.14}$$

for the transmitted wave the interfacial boundary condition gives

$$R = (\nu_0 - \nu_1)/(\nu_0 + \nu_1)$$

and

$$T = 2\nu_0/(\nu_0 + \nu_1) \tag{5.15}$$

which can be expressed concisely in terms of the parameter

$$X = (N_1^2 - N_0^2)/\nu_0^2(U-c)^2 \tag{5.16}$$

$$R = \left(1 - (1+X)^{\frac{1}{2}}\right)/\left(1 + (1+X)^{\frac{1}{2}}\right)$$
$$T = 2/\left(1 + (1+X)^{\frac{1}{2}}\right) \tag{5.17}$$

For $X \gg 1$, the second medium is so highly stratified that it is capable of transmitting all species of waves, but it is also very resilient in the sense that almost all the energy is reflected. We interpret this in terms of the impedance at the interface not being matched. The intermediate case $X = 1$ gives $T = 0.83, R = -0.17$. From these figures we get the impression that, though perfect reflection is the limit case, it is a rather extreme one, and a layer of increased stability generally implies propagation with some reflection. Notice that the wave is refracted towards the vertical in the more stable medium.

For $X = 0$ there is no change in the medium, and consistently all the energy is transmitted.

For $-1 < X < 0$ the second medium is reluctant to allow, but is not incapable of, transmission of the wave, $R \simeq 1, 2 > T > 1$, and the wave is refracted towards the horizontal. The limit case $X = -1$ can be visualised as perfect reflection of the incident wave, which also penetrates a great distance into the upper medium.

For $X < -1$, v_1 is purely imaginary and all the energy is reflected.

That the amplitude of the transmitted wave is sometimes greater than that of the incident should cause us some concern. However, a little thought tells us that we should expect restrictions on energy propagation rather than on the amplitude of the waves.

5.4 Propagation of energy

We can define the components of the group velocity through the concept of stationary phase as $(\partial/\partial k, \partial/\partial v)(kc)$, where the phase speed c is relative to the ground. This is rather arbitrary unless we are specifically interested in propagation relative to the ground, as we might be in the situation where the waves are generated by a stationary obstacle such as a mountain. For some applications phase and group velocities measured relative to the air parcel are more useful. To do this we can use a Lagrangian system, replacing c by $c - U$, which implies a travelling coordinate replacing x by $x - Ut$. In this system we have, for gravity waves phase velocity

$$(1, k/v) \, (c - U) \tag{5.18}$$

and group velocity

$$(v^2, -vk) \left(\frac{(c - U)}{(k^2 + v^2)} \right) \tag{5.19}$$

We notice that the vertical component of phase and group velocities are in opposite directions, and that a layer for which $v = 0$ cannot transmit energy. Solutions valid to order ϵ can be used to calculate terms of order ϵ^2, with error of order ϵ^3. We can then check from the solution that the terms in $\frac{1}{2}v^2$ and $\frac{1}{2}g\,(\delta\phi)^2$ averaged over one horizontal wavelength, or over one temporal period, are equal; not surprising since the wave represents repeated cyclic exchange between these two forms of energy.

For a plane wave of form

$$\psi = a \exp \mathrm{i}(k(x - ct) + vz)$$

the energy density is

$$\rho_0(v^2 + k^2)| \, \psi \, |^2 \tag{5.20}$$

and the energy flux is

$$\rho_0 vk \, (U - c)| \, \psi \, |^2 \tag{5.21}$$

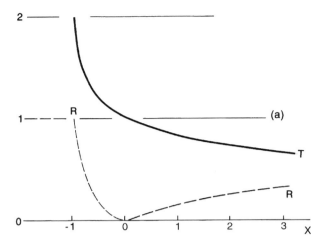

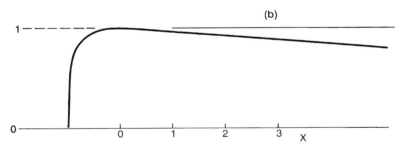

Figure 5.2 (a) Amplitude of the transmitted gravity wave T and amplitude of the reflected wave R, as a function of the stratification parameter X defined by equation 5.16. (b) Energy flux carried by each component. All now have maximum value unity, and add to unity throughout the range. Notice the rather abrupt change between almost perfect transmission and almost perfect reflection.

This expression can be used to show that the solution set defined by equation 5.19 has the same flux of energy towards the interface as away. The physical mechanism that transfers energy between these inviscid layers is the work done by the pressure at their boundaries. From equation 5.7 we see that the vertical flux of energy into a parcel is $\overline{w'\,\delta p'}$. We deduce $\delta p'$ from the horizontal component of momentum, the linearised version of equation 5.1, which gives

$$\partial p'/\partial x = -\rho(U - c)\partial u'/\partial x - w'\,dU/dz \qquad (5.22)$$

for a wave moving with velocity c in the x-direction, and not otherwise restricting the variation of $B(z)$ or $U(z)$. The second term of equation 5.22 does not contribute to the flux because it is out of phase in the x-direction. Thus, rather generally the vertical flux is $-\rho(U-c)\overline{u'w'}$ which agrees with equation 5.21 when the variation of U and B can be neglected. We notice that the energy flux is associated with the momentum of the fluid relative to the wave. The energy fluxes can again be expressed in terms of the parameter X of equation 5.16, and gives the lower set of curves in Figure 5.2.

5.5 Gravity waves generated by stationary flow over an obstacle

Suppose that the obstacle is defined by the profile $z = h(x)$, then the linearised condition that the flow follows the ground is $w = Dh/Dt$ at $z = h$. The linearised version of this condition is

$$w = U\ \partial h/\partial x \text{ or } \psi' = -Uh \text{ at } z = 0 \tag{5.23}$$

which is the linearised version of making the stream function constant at the ground, and is a much nicer way of expressing the boundary condition. Consider stationary flow, defined by $\partial/\partial t = 0$, and also suppose U and B to be independent of height. For each Fourier component of h of form $a\exp ikx$, the corresponding stream function is

$$\psi` = -aU\ \exp \mathrm{i}(kx \pm vz) \text{ where } k^2 + v^2 = N^2/U^2 \tag{5.24}$$

There are two ranges of solution. For waves so short that $k^2 > N^2/U^2$, v is purely imaginary, and ψ' varies exponentially in the vertical. If the fluid is sufficiently deep, the solution that increases upwards will be excluded giving

$$\psi' = -aU\ \exp (\mathrm{i}kx - \mid v \mid z) \tag{5.25}$$

which resemble the irrotational solutions of classical fluid mechanics, to which they reduce for very short horizontal wavelengths. The flow is symmetrical relative to the crests of the ridges, so there is no net pressure force on the ground. Disturbance is confined to a height of order $1/\mid v \mid$ above the obstacle, and so penetrates further as the horizontal wavelength increases, until that wavelength reaches the critical value $k^2 = N^2/U^2$. For horizontal wavelengths longer than this, v is purely real, ψ' behaves like $\exp \mathrm{i}(kx + vz)$ or $\exp \mathrm{i}(kx - vz)$ or a linear combination of both. If there is no reflecting layer then the boundary condition is one of no downward propagation of energy. This eliminates the solution with $-\mathrm{i}vz$ and leaves

$$\psi' = -aU\exp \mathrm{i}(kx + vz) \quad \text{for which } \delta p' = \mathrm{i}\,\rho a v U^2 \exp \mathrm{i}(kx + vz) \tag{5.26}$$

Recollect that it is the real part of these expressions that is the physical solution; it is necessary only that the differential equations be linear, and their coefficients real. We find that the pressure is $\pi/2$ out of phase with the height with minimum pressure on the downwind side, maximum on the upwind side. The average net force acting on the ground is

$$\frac{1}{L}\int_0^L \delta p' \frac{\partial h}{\partial x}\, \mathrm{d}x = \frac{1}{2}\rho a^2 k v U^2 = \frac{1}{2}\rho a^2 k\left(\frac{N^2}{U^2} - k^2\right)^{1/2} U^2 \qquad (5.27)$$

This can be compared with the aerodynamic or Newtonian drag law; $drag = \rho C_D U^2$. Values of *several* 10^{-3} are usually quoted for C_D for rough ground, corresponding to drag of about $0.1\,\mathrm{N\,m^2}$ for winds of $10\,\mathrm{m\,s^{-1}}$. If we neglect the variation of a with k, then equation 5.27 has maximum contribution to drag for $k \simeq 0.7 N/U$ which gives

$$drag \simeq 0.25\,\rho a^2 N^2 \qquad (5.28)$$

This value of k corresponds to a wavelength of some $10\,\mathrm{km}$ at which we would expect $a \simeq 10$ to $100\,\mathrm{m}$ over land. Thus, with $N \simeq 10^{-2}\,\mathrm{s^{-1}}$ this mechanism may just contribute significantly to the surface drag.

Some points of principle are raised. So long as the flow is not rapidly changed by the drag, stress can be equated with momentum flux. We check from the solution that the momentum flux $\rho\,\overline{u'w'}$ is equal to the drag due to the pressure.

The momentum extracted at the ground is taken out of the layer of air above, indeed through the same pressure force as at the solid surface below. Thus our layer of constant U and N serves to transmit the stress from the surface air up to some upper level, maybe of the order of one kilometre above. This is very different to the effect of the usual transmission of momentum from layer to layer carried by turbulence that derives energy from the shear. In that case momentum is taken from a layer usually somewhat shallower than $100\,\mathrm{m}$ in depth, next to the ground. This mechanically driven turbulence can therefore slow down the surface flow, relatively easily. We see this in the 'logarithmic' boundary layer where the momentum equation reduces to a diagnostic equation for the wind profile. In contrast, the gravity-wave drag is transmitted upwards, does not slow down the layer through which it is transmitted, but some other.

From what level does the momentum transferred by the gravity wave come? Perhaps some sort of dissipation causes the amplitude of the wave to decrease with height in some layer. The wave then cannot carry as much momentum as it did at the lower level, and it must be the flow in this dissipation layer which is slowed down.

Perhaps the wave becomes so distorted as to find itself in an irreversible situation, wave breaking, and so-called critical level absorption, where $U \simeq c$ over a considerable depth, are possibilities. Such a situation might be detected as a great restriction on the amplitude of solution that could still be regarded as satisfying a criterion for linearity, or as vulnerability to temporal instability.

The wave might get lost. If there was a perfectly reflecting layer at some finite height, then the steady solution would have equal contributions from the terms in $\pm \nu$ and there would be zero drag. However, seen as an initial value problem, the wave has to travel to the upper reflector before it can be relected to bring the wave represented by the opposite sign of ν back. If the larger-scale flow has changed by the time the wave returned from the reflecting layer, this would mean loss of the phase coherence which is needed to exactly cancel out the initial drag created when the wave first set out. This idea appeals to me because it has a degree of novelty, suggests an underlying uncertainty on the response of the atmosphere to what seems superficially to be similar situations (unpredictability!) and suggests some feedback between the large-scale and small-scale flow.

That the gravity-wave drag predicted by equation 5.28 is independent of U is of some interest. This curious relation perhaps makes us think of air being dammed up behind obstacles as in Figure 5.3, and the horizontal force needed to maintain the slope. In the Boussinesq system the force/unit area, for denser fluid leaning on obstacles of height a, separated by distance d is

$$\frac{g\, \delta \rho'\, a}{d} \simeq \frac{\rho d\ \delta \phi\, a}{d} \simeq \rho(\tfrac{1}{2}Ba)\frac{ga}{d} \qquad (5.29)$$

rather like the limiting case of equation 5.28, for the drag of gravity waves. Something like this may happen on those occasions when the surface air cools violently by intense electromagnetic-radiation cooling near the ground at night, and the surface wind seems to vanish completely. Something must balance the large-scale pressure gradient, and perhaps it is this damming process.

In general there is a spectrum of orography, and the surface drag will contain contributions from the nearly hydrostatic effect of several different scales of orography each with $k^2 \ll N^2/U^2$, and dependent on velocity, as well as the non-hydrostatic contribution.

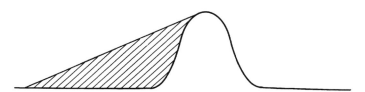

Figure 5.3 Air dammed up behind an irregularity in the ground giving a net horizontal force in spite of zero surface wind. This picture is not inconsistent with the limiting case of proagating gravity waves as discussed in Section 5.5. Can the surface wind vanish on such a night, in spite of a pressure field on the larger scale?

5.6 Gravity waves generated by an isolated obstacle

Given a more general obstacle $h(x)$, the linear flow over it follows from the Fourier integral representation

$$h(x) = \int_0^\infty f(k) \exp \mathrm{i}kx \, \mathrm{d}k$$

whence

$$\psi' = -U \int_0^\infty f(k) \exp \mathrm{i}(kx + vz) \, \mathrm{d}k \qquad (5.30)$$

where $v^2 = N^2/U^2 - k^2$, and v is positive real, or of positive imaginary part in appropriate ranges of k. Stationary phase now properly relative to the ground, is where

$$(\mathrm{d}/\mathrm{d}k)(kx + vz) = 0$$

which gives

$$k = k_0 = v_0 x/z \quad \text{and} \quad k_0^2 + v_0^2 = N^2/U^2 \qquad (5.31)$$

where we suppose that the function $f(k)$ is fairly smooth; implying that we are at least several half-widths away from the obstacle in the horizontal before stationary phase becomes a useful concept. We see that the condition of stationarity depends on position in space; much like it does for optical interference behind a small number of narrow slits. We notice that the condition of no downward flux of energy demands that v must be real and positive, so equation 5.31 shows that constructive stationary phase exists only downwind of the obstacle.

Introducing the scaled radial distance $r = N(x^2 + z^2)^{1/2}/U$ and expanding the wavenumber k about the point of stationary phase gives

$$kx + vz = r - \theta^2 - (x/z)(2/r)^{1/2}\theta^3 + (O\theta^4) \qquad (5.32)$$

where we have introduced the new non-dimensional wavenumber variable $\theta = (r/2)^{1/2}(k - k_0)/v_0$, which is of size unity where the integrand is large, in order to test for neglect of terms in the expansion. We see that the term in θ^3 can be neglected if we are far enough away from the obstacle, but not too low. In this zone the integral equation 5.30, can be expanded to give

$$\psi' = -\sqrt{2} f(k_0) \frac{N^2 z \, \mathrm{e}^{\mathrm{i}r}}{U r^{3/2}} \int_{-\theta_0}^\infty \mathrm{e}^{\mathrm{i}\theta^2} \, \mathrm{d}\theta$$

where $\qquad\qquad\qquad\qquad\qquad\qquad\qquad\qquad\qquad\qquad (5.33)$

$$\theta_0 = \frac{x}{z}\left(\frac{r}{2}\right)^{1/2}$$

Now we identify the sub-range $\theta_0 \ll 1$, in the quadrant above and downstream , for which

$$\psi' \simeq -(2\pi)^{1/2} f(k_0) \ (N^2 z/U) r^{-3/2} \exp \mathrm{i}(r - \pi/4) \tag{5.34}$$

and the small section $\theta_0 \ll 1$, in the thin slice directly above the mountain, where ψ' is about half this value. Further juggling shows that far upstream/downstream, where $z \ll |x|$ and $|Nx/U| \gg 1$, the dominant part of the integral is from a path along the negative/positive imaginary axis of k, which gives

$$\psi' \simeq -\mathrm{i} f(0) (U/x) \exp \mathrm{i}(Nz/U) \tag{5.35}$$

Figure 5.4 shows one solution calculated by numerically integrating the Fourier integral. Using our analytic solution we see:

(1) A quadrant of constructively interfering waves extending above and downstream of the obstacle; amplitude decreasing like $r^{-1/2}$ which gives finite energy propagation into the quadrant.

(2) A region of destructive interference in the upstream quadrant, where the amplitude decreases as $1/x$. These solutions merge over the crest with the constructive solution (1). The horizontal velocity near the ground is decreased upstream, and that downstream is increased, according to equation 5.35, like $f(0) N/x$ so the flow upstream tends to be blocked. Now $f(0) \sim ah$ where a is the

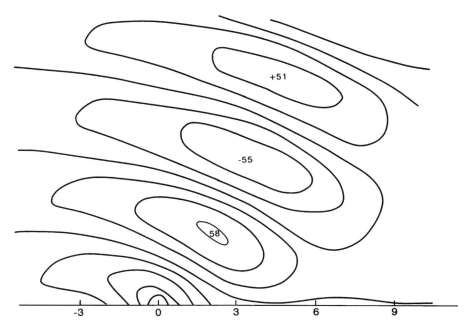

Figure 5.4 Perturbation stream function in the x–z plane for flow with velocity and stratification independent of height crossing a 2-D ridge. We see waves propagating into the quadrant downwind of the ridge decaying slowly, and waves interfering upstream to give fluctuations varying in the vertical direction, flow slowest near the ground, decaying quickly upstream.

halfwidth of the ridge and h is the height, and one might speculate that for $hN \sim U$ there might be complete blocking.

(3) A shadow region separating the constructive region from the ground near the lower boundary downstream. We might be astonished that such complexity arises from the solution to the simple equation $(\partial^2/\partial x^2 + \partial^2/\partial z^2 + 1)\psi = 0$ with a simple point source and simple radiation boundary conditions, but we should compare the solution with the optical analogy of diffraction through a slit, when we are concerned with describing the nature of the wave near the slit edge, and the shadow zone, on the scale of the wavelength of the wave.

5.7 Trapped waves (lids lead to resonance)

had we but time enough and space

For some wavelengths, energy can be transmitted upwards in a lower layer but not through an upper layer. A very explicit example is when the vertical velocity vanishes at some finite height which is independent of wavelength; the rigid lid condition. The solution is of the form

$$\psi = A \exp ikx \ \sin \nu(z - z_A) \tag{5.36}$$

where the lid is at height z_A. The boundary condition at the ground gives

$$\psi = U \int_0^\infty f(k) \exp ikx \frac{\sin \nu(z - z_A)}{\sin \nu z_A} \, \mathrm{d}k \tag{5.37}$$

in place of equation 5.30. The integral now has the possibility of poles (singularities contributing finitely to the value of the integral) where the denominator vanishes. These are at $\nu z_A = 0, \pi, 2\pi, \ldots$. We notice that these are resonant waves in the sense that they are solutions to the unforced system having zero vertical velocity at the top and bottom boundaries. We can visualise such waves as starting out from the ground, being reflected off the upper lid and arriving back at the ground in phase with the wave just starting out. Perhaps for $\nu z_A = \pi$ the wave returns to reinforce the next wave, for $\nu z_A = 2\pi$ the next but one, and so on.

For $\nu z_A = 0$ both numerator and denominator vanish, so this is not a pole. The first pole is for $\nu z_A = \pi$ $k = k_1$ where $k_1^2 = (N/U)^2 - (\pi/z_A)^2$ and the contribution to the integral is

$$i\pi^2 \left(\frac{U}{z_A^2 k_1} \right) \exp ik_1 x \ \sin \left(\frac{\pi(z - z_A)}{z_A} \right) \tag{5.38}$$

Instead of spreading out radially into the quadrant behind the obstacle, and consequently decreasing in amplitude like $r^{-1/2}$, the wave is trapped in the channel $0 < z < z_A$, and the amplitude does not decrease with distance at all.

Well perhaps it spreads out and decreases in amplitude until it fills the channel, then attains constant amplitude.

Cloud bands, associated with such trapped waves, can often be seen, particularly downwind of mountainous areas of the Earth. They are also seen in the atmosphere of Mars, where they provide interesting information. It is one of the challenges typical of remote sensing, that we can often detect obscure combinations of individually useful variables. In this case we can observe the combination $N^2/U^2 - \pi^2/z_A^2$, but would prefer to have had more universal quantities like U or N or z_A separately.

For real hills the flow is around, as well as over. This raises considerable problems for the mathematical analysis, even for merely visualising the solutions. Trapped waves now spread out horizontally, as well as possibly vertically, giving a picture much like the disturbance caused by a ship travelling through the sea, with cusped waves at the edge of the region of spreading lee waves. Further study concerns the behaviour of flow near the ground, where the tendency noted for the stronger winds and lower pressures to occur on the downwind slopes is verified. We begin to see that the flow round hillocks on hills may also be a scaled down version of flow round hills.

If the obstacle slopes steeply, the flow may not follow the contour but break away, and give a recirculating eddy. Such flow can have a marked effect on picnic sites, and other mechanisms for concentrating litter, as well as being good places for finding insects and soaring birds. So far as the flow of larger scale (like our lee waves) is concerned, these circulations of smaller scale serve mainly to fill in the hollows in the ground and smooth the orographic profile. Frictionally driven turbulence causes modification of the wind profile which may be significant for the siting of trees and wind-driven machinery.

Chapter 6

Shearing instability

6.1 Helmholtz instability

When variations in velocity are very important, but stratification less so, we find that the flow is often unstable. If the flow is incompressible, unstratified, and two-dimensional, then equation 5.4 shows that vorticity is not generated in the flow, but only moved about by it. Now suppose that all the initial vorticity is concentrated in a very narrow region of space in that the tangential velocity U changes from one relatively constant value to another in a small height range centred on $z = 0$.

We idealise the situation by setting

$$U = U_1 \text{ for } z > h$$
$$= U_2 \text{ for } z < h$$

where $z = h(x, t)$ is the position of the interface between the two streams. We see that there is a vortex sheet at the interface. Everywhere else the flow is irrotational so $\partial^2\psi/\partial x^2 + \partial^2\psi/\partial z^2 = 0$. For wavelike perturbations such that $\psi \propto \exp ikx$, then

$$\psi = A_1 \exp -kz \text{ for } z > h$$
$$= A_2 \exp +kz \text{ for } z < h$$

That the interface moves with the fluid next to it demands $D(z - h)/Dt = 0$, from both sides of the interface. With $h = \alpha \exp ik(x - ct)$, this gives

$$(U_1 - c)\alpha + A_1 = 0$$
$$(U_2 - c)\alpha + A_2 = 0 \qquad (6.1)$$

so long as the perturbation is of small enough amplitude, i.e. neglecting all non-linear terms. Continuity of pressure at the interface demands that $\partial p/\partial x$ (from the linearised x-component of the momentum equation) be the same from both sides, or

$$(U_1 - c)(-kA_1) = (U_2 - c)(kA_2) \qquad (6.2)$$

and this set of three equations has solutions not identically zero only if

$$(U_1 - c)^2 + (U_2 - c)^2 = 0$$

giving

$$c = \frac{1}{2}(U_1 + U_2) \pm i\frac{1}{2}(U_1 - U_2) \qquad (6.3)$$

The real part of c shows that the perturbation moves with the average velocity of the unperturbed flow. The imaginary part c_i shows that the amplitude of one of the solutions increases exponentially with time. The amplification rate is kc_i showing that shorter waves grow faster.

The time taken for a particle to pass through the wave is $(2\pi/k) / \frac{1}{2}(U_1 - U_2)$. In this time interval the amplitude of the wave will have increased by a factor of $\exp 2\pi \sim 530$ so we see that the fluid moves only a little way through the wave while the amplitude of the wave increases greatly. In less idealised flows, like those of Section 6.4 for example, the growth is less extreme and the amplification factor may be more like 12, so the proposition remains true but muted. This comparison of time scales is arbitrary, and later we notice that, when we need timescales for comparing temporal derivatives, we should use the temporal frequency U/k not the (inverse) period $2\pi U/k$. We notice that neither the tangential, nor the normal components of velocity are continuous at the interface.

To picture the motion we examine one of the growing solutions. Without loss of generality we have chosen a time such that; $\text{Real}(\alpha \exp -kc_i t) = -\epsilon$ with ϵ real. For example we have, at $z = 0$

$$h = -\epsilon \cos kx \qquad (6.4)$$
$$u_1 - u_2 = -2\epsilon \sin kx \qquad (6.5)$$

and

$$p' = \frac{1}{2}\rho_0 k\epsilon (U_1 - U_2)^2 \cos kx \qquad (6.6)$$

where we have chosen axes that move with the mean flow, by making $U_1 + U_2 = 0$. Figure 6.1 shows some aspects of the solution.

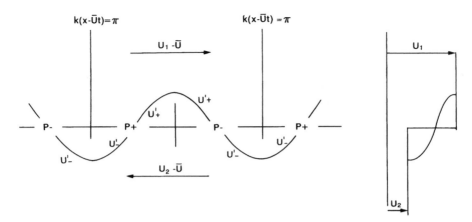

Figure 6.1 Helmholtz idealisation of shearing instability where flow is in opposite directions above and below an insubstantial interface. This interface moves as the instability develops with sinusoidal ripples increasing in amplitude exponentially with time. We then see penetration of one layer by the other and momentum exchange by the pressure forces, decrease in the vorticity contrast as the interface becomes longer, but no mixing. The x-average velocity field is shown on the right.

The vortex sheet has begun to wrinkle up with fluid from each side penetrating the other. This implies that the x-average flow has been smoothed out over the region $-\epsilon < z < \epsilon$, as shown on the right side of Figure 6.1. Pressure at the interface is larger on the 'upwind' sides of the slopes in both layers. Thus momentum is transferred between the fluids by pressure forces. Both of these are evidence that kinetic energy of the mean flow is being converted into kinetic energy of the eddy motion; inevitable, since there is no other source of kinetic energy for the amplifying perturbation. We notice that the transfer of momentum is independent of mixing in any sense.

The vortex sheet becomes thicker in some places and thinner in others. This asymmetry takes us towards the classical shape of a breaking wave, as in Figure 6.2, with the tendency towards the penetration and wrapping up of one fluid in the other. The vortex sheet gets longer, but total vorticity must be conserved, so the velocity discontinuity must get generally weaker as the vortex sheet becomes longer.

6.2 Helmholtz waves of finite amplitude

As the vortex sheet begins to distort, various non-linear effects become apparent. Following the behaviour of the sheet in detail is rather difficult, but we can make an interesting model of the process. Instead of following a true discontinuity in velocity we can suppose that there is a row of rather many, rather weak,

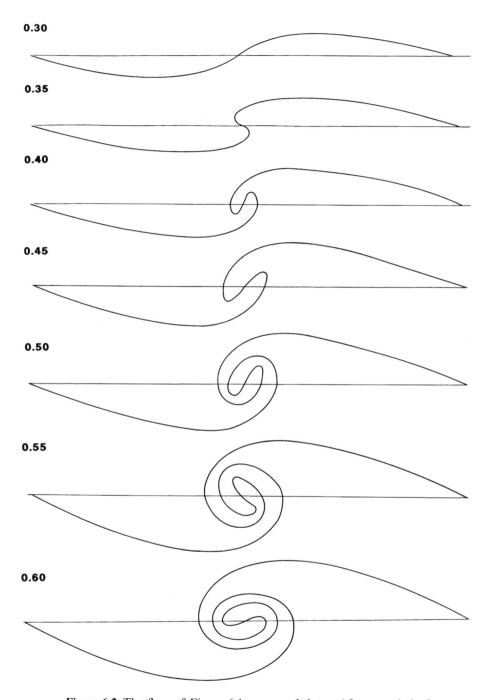

Figure 6.2 The flow of Figure 6.1 seen much later. After a period of exponential growth the motion seems to settle down to the task of winding up each layer inside the other as 'Swiss roll' turbulence. What happens at similar narrow regions in a real diffusive system is explored a little more in Chapter 14.

point vortices. In the limit of large number this would give a simple discontinuity in tangential velocity, but we can model this by just keeping a few vortices
(which then do not represent a perfect discontinuity in velocity) and follow the
path of each vortex as it is advected by the others. Because the initial perturbation is sinusoidal in space the motion will continue to be periodic, though not
sinusoidal, in space.

For convenience we scale the wavelength to unity. The influence of vortices
displaced by unit wavelength can be taken into account rather readily. Velocity
induced at the point (x', y') by a set of vortices at positions (x, y), $(x, y + 1)$,
$(x, y - 1)$, $(x, y + 2) \ldots$ etc. is readily seen to be

$$V_x = -\pi S \sinh 2b \qquad V_y = \pi S \sin 2a$$

where

$$S = 1/(\cosh 2b - \cos 2a), \quad a = \pi(x' - x), \ b = \pi(y' - y) \qquad (6.7)$$

Integration of even such a simple system is not without interest. With only two
vortices, one initially close to $x = 1/4$, the other initially close to $3/4$, there
exists an integral

$$\cosh 4\pi y - \cos 4\pi x = constant \qquad (6.8)$$

showing that the vortices cannot move indefinitely far from the x-axis. It then
becomes more-or-less obvious, that the solutions in this simple case must be
periodic in character. In fact they are rather like elliptic integrals, but the
numerical evaluation of the integrals is sufficiently complicated that one
might as well solve equations for the path numerically, as they stand.

The small-amplitude wave with two repeated vortices has infinite period;
starting exponentially with time from one unstable equilibrium state and
approaching exponentially with time, the other; just like a rigid pendulum displaced slightly from a vertically unstable equilibrium position.

Three vortices seem to be more chaotic, with a tendency to spend more time
in some regions of space which moves with the wave. With more vortices we
begin to approach a recognisable vortex sheet, and are tempted to join their
positions with a continuous line which produces an evolution like that shown in
Figure 6.2. As the motion proceeds, the interface lengthens as shown in Figure
6.3, and the fluids on each side have more area for potential diffusion of inert
properties but not vorticity.

This looks very much like an idealistic interweaving situation; what I like to
call 'Swiss Roll' turbulence after the English confectionary consisting of a sheet
of sponge with a sheet of jam spread on the top, with a sheet of cream on top of
that, which are then rolled up together.

Remember however that the growth rate of displacements is, in the linear
case at least, proportional to the separation of the vortices. For the case with 16
vortices this means that the amplitude of the irregularities between adjacent

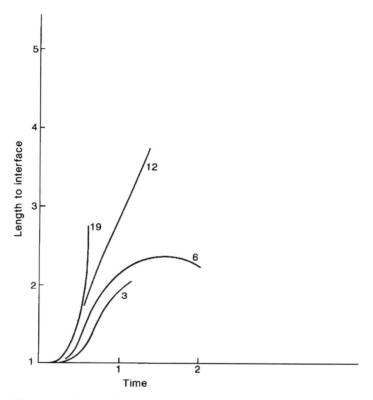

Figure 6.3 Elongation and weakening of the vorticity sheet at the interface region is one interesting parameter of shearing instability. The spatial resolution of the model is given by n: the number of point vortices in one wavelength. As the spatial resolution increases from 3 to 6 to 12 to 19 the interface seems to lengthen more violently. Thus while Figure 6.2 shows inexorable winding up, higher resolution shows a similar picture happening earlier on. It may be that there is no solution to the inviscid problem but it is difficult to disentangle numerics from physics. As the resolution is increased further to 29 the limit to which the computation can be trusted begins to get shorter because the truncation errors in the high wavenumbers grow more quickly. It is a nice philosophical problem whether a well-defined solution exists. It is a nice computational problem to find such a solution, assuming it does.

vortices is expected to amplify 16 times faster than the basic wave. Indeed, what is an irregularity in this context? Should we, for example, expect the harmonics of the basic wave to grow away from the initial smooth wave with such explosive rapidity? Numerical integrations suggest that the solution which is of constant shape when the amplitude is small, will grow smoothly in the manner shown in Figure 6.2 for a time but that the solution eventually becomes distorted by the shorter wavelengths, amplifying at about the rate expected by the linear analysis.

Now, the further back in time that the integration starts, and the more resolution one employs, the more time there is for amplification of more-virulent spurious waves and the worse this problem becomes. The situation is to some extent artificial because the initial state contains all wavelengths of which the shortest are the most violent, but it does highlight one aspect of the problem of predictability. Another way of seeing the artificiality is to recognise that, because the point vortices have no spatial extent, they can be packed indefinitely into a small region. Let us speculate on the possible influence of viscosity.

6.3 Short waves and viscosity

We notice that in the linear solution, the shortest waves grow most quickly, and imagine that viscosity might serve to limit their growth. This raises something of a paradox because, as shown in equation 3.25, apart from the effect of boundaries which are here supposed remote and ineffective, viscosity serves to smooth out only the field of vorticity, so has no effect on irrotational flow which occupies most of the space in the Helmholtz problem. In addition, even the unperturbed flow is not consistent with the existence of viscosity, in the sense that we cannot have an unbounded shear, represented by a discontinuity in velocity, in a viscous medium.

In spite of these misgivings, we might estimate that the rate of damping of wave motion due to viscous dissipation with kinematic coefficient of viscosity v_v, would be about $v_v k^2$. This we might compare with a shear-driven growth rate of $k \Delta U$. Thus there might be a viscous limit to growth where

$$v_v k^2 \sim k \Delta U \qquad (6.9)$$

But what value should we use for ΔU ? Arbitrarily putting $\Delta U = 1 \, \text{m s}^{-1}$ gives $2\pi/k = 0.02 \, \text{cm}$, clearly an unreasonable conclusion because we do not get such large changes in velocity over such small distances. We can avoid this defect by setting $\Delta U \simeq (1/k)\partial U/\partial z$, so that the velocity difference driving the eddy is related to the scale of the eddy. The resulting length scale

$$1/k^* = (v_v/\partial U/\partial z)^{1/2} \qquad (6.10)$$

is recognised as roughly separating an inertial, or energy-transfer range of turbulent breakdown from the range of viscous dissipation. For example applying this in the logarithmic boundary layer gives for $\partial U/\partial z \sim 0.1 \, \text{s}^{-1}$, $1/k^* \sim 2 \, \text{cm}$ which seems about right. Notice that we take almost nothing from our current mathematical analysis to arrive at this argument; we could have done it all from first principles.

6.4 Short waves; distributed shear

Consider another hypothesis for determining a non-zero scale for shearing instability. Following Rayleigh, we consider a velocity profile with a finite region $0 < z < h_0$ of constant shear. The perturbed motion is irrotational in all three layers, because there are no variations of vorticity to be advected about in any of the layers. The calculations are straightforward; stream function and normal velocity are both continuous at the interfaces, whose disturbed position is not important in the linear problem, and the calculations give

$$kc = U(1 - 2kh_0 + k^2h_0^2 - \exp{-2kh_0})/2kh_0 \tag{6.11}$$

whose solution is shown in Figure 6.4. For waves so short that $kh_0 \gg 1$

$$c = \pm\frac{1}{2}\Delta U\left(1 - 1/kh_0 + O(kh_0)^2)\right) \tag{6.12}$$

from which we deduce that such short waves are unable to detect the velocity difference and cannot convert kinetic energy of the mean flow into kinetic energy of the eddy. We notice that the eddy flow penetrates in the z-direction

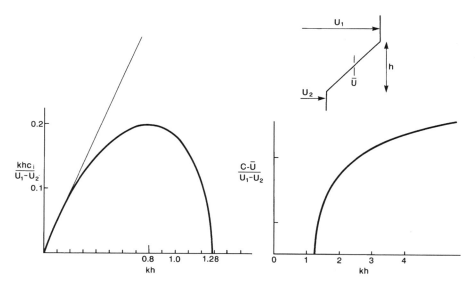

Figure 6.4 Variation of amplification rate and phase speed with wavenumber for Rayleigh's idealisation of shearing instability where the velocity transition occupies a layer of non-zero thickness. To waves that are much longer than the width of the interface the Helmholtz (thin line) and Rayleigh (thick line) solutions are similar because the waves are too long to be able to detect the width of the interface region. Short waves do not grow because they cannot detect the difference in velocity across the sheared layer. However, as the interface lengthens and narrows it is likely to become unstable to shorter waves so the problem of a deterministic scale for the flow may only be delayed.

like exp $-kz$, which means that they can detect only the change in shear at one edge of the shear zone, which is not enough to provide eddy kinetic energy. The short waves do not amplify but only get blown along by the wind at the edge of the shear zone. We wonder why.

For very long waves, with kh_0 small, the solution reverts to that of the Helmholtz problem (because the motion cannot detect the subtleties of the gradual shear) and one solution amplifies exponentially.

Maximum growth rate is $kc_i \simeq 0.20 \Delta U/h_0$ at a wavelength such that $kh_0 \simeq 0.80$, and there is stability for $kh_0 > 1.28$ as illustrated in Figure 6.4. But having solved this more general instability problem, and deduced a non-zero characteristic scale for instability, are we really much further forward in understanding the problem of mechanically driven turbulence, for now we have had to introduce the space-scale h_0 which surely must be expected to arise from more fundamental considerations. We wonder if the finiteness of the shear zone might be the result of shearing instability in an earlier, more nearly discontinuous velocity profile. As L.F.Richardson put it;

> Big whirls have little whirls
> that feed on their velocity,
> and little whirls have littler whirls
> and so on to viscosity.

Perhaps we should consider the molecular diffusion of momentum between the two initially discontinuous streams at the same time as dynamical instability sets in? We might try to pose a more general turbulence problem by supposing that the input of energy was at a given, fairly large scale and that there was explicit viscosity at a much smaller scale. I tried to do this in a two-dimensional problem, using an input of momentum that had a well-defined space scale because it was sinusoidal in the transverse direction. The sinusoidal flow that results is unstable if the effect of viscosity is not overpowering; there is a critical Reynolds number. To my surprise, the development of the instability into the non-linear regime leads to a new steady state, in which steady eddies transfer momentum steadily between source and sink. I find this astonishing.

Also remarkable, though perhaps not astonishing, is that the resulting steady eddy flow is on a greater scale than that with which we started. Thus as the eddy develops from small amplitude, there is a distribution of energy to longer, as well as to shorter, space scales. This is because the mean-square of both vorticity and kinetic energy are conserved by the inviscid motion. Energy at the shorter scales is then dissipated by viscosity, leaving the remaining energy at the longer scales.

Because of this increase in scale, the new steady-eddy state might be even more unstable than the original flow. This suggests that for some flows instability may be a way of making the transition from one steady flow to another, while in others it may lead to catastrophe. Alternatively, the turbulence might be breaking out, smoothing the original shear, then calming down; a sort of intermittent action. An interesting case concerns the generation of vortex streets behind a bluff obstacle. It is not unlikely that the obstacle generates two sheared layers through the viscous interaction with its own boundaries, and that these are shearing-unstable, so break up into concentrated vortices. That these vortices are alternate on each side would seem not unreasonable from general energetic considerations, and we have an idealisation of Karman vortex streets.

6.5 Short waves and stratification

If the original flow is stably stratified as well as sheared, some reduction in the rate of amplification can be expected. Thus particles are reluctant to move up and down because of the work they must do against gravity, and gravity waves may propagate energy away from the shear zone. If both velocity and density are discontinuous at $z = 0$, and constant everywhere else, then only the work done against vertical displacement is represented, and a fairly simple extension of the Helmholtz analysis, in which $\partial p/\partial z = -\rho g$ must be included in the consideration of continuity of pressure at the interface gives

$$(\rho_1 + \rho_2)c = \rho_1 U_1 + \rho_2 U_2 + \left((g/k)(\rho_1^2 - \rho_2^2) - \rho_1\rho_2(U_1 - U_2)^2\right)^{1/2} \tag{6.13}$$

Amplifying waves exist if

$$(U_1 - U_2)^2 > g\,(\rho_1^2 - \rho_2^2)/\,\rho_1\rho_2 k \tag{6.14}$$

Short waves ignore the stratification, astonishingly whether that stratification is gravitationally stable or unstable. When the density change is small, amplifying waves exist if

$$g\frac{(\rho_1 - \rho_2)}{\rho} > \frac{1}{2}\,k\,(U_1 - U_2)^2 \tag{6.15}$$

If the shear is distributed over distance h_0 say, then we might suppose, as in Section 6.4, that the most unstable wave was such that $kh_0 \simeq 1.3$, and if so, criterion 6.15 becomes

$$g(\rho_2 - \rho_1)/\rho > 0.65(U_1 - U_2)^2/h_0 \tag{6.16}$$

nearly. We recognise the Richardson number, and have the rough criterion $Ri > 0.65$ for stability.

6.6 Energetics of sheared stratified overturning

The critical Richardson number for stability probably depends on the shape of
the density and velocity profiles but 1/4 seems a well-established value. A fairly
general argument supports this. Consider a shallow layer in which the shear and
the static stability are constant. If no work is done from the outside, the total
kinetic plus potential energy for the layer must remain the same, as must the
total momentum and thermal energy.

Relative to a suitable coordinate system, the original state can be written as

$$U = \alpha z \qquad \phi = \phi_0 + B z \text{ for } -h < z < +h \tag{6.17}$$

If the final state is one of complete mixing, then it will be such that

$$U = 0, \quad \phi = \phi_0 \tag{6.18}$$

over this same region. We can now calculate the kinetic and potential energies
of the initial and final mean states

$$\text{initial K.E.} = \int_{-h}^{+h} \frac{1}{2}\rho(\alpha z)^2 \, dz = \frac{1}{3}\rho h^3 \alpha^2 \tag{6.19}$$

$$\text{initial P.E.} = -\int_{-h}^{+h} \rho g z \, B z \, dz = -\frac{2}{3}\rho g B h^3 \tag{6.20}$$

We have chosen a coordinate system such that the final potential and kinetic
energies of the mean flow are each zero, so the sum of the initial potential and
kinetic energies might be available to make eddy kinetic energy. This amounts
to

$$\rho h^3 (\alpha^2 - 2gB)/3 \tag{6.21}$$

Thus, according to equation 6.21, there may be energy available to do the
mixing so long as $Ri = (gB/\alpha^2) < 1/2$. Curiously this is *not* the criterion
$Ri < 1/4$ which we have just quoted for the onset of shearing instability.

Then we notice that the theoretical developments suggest that momentum
will be transferred rather efficiently by pressure forces between streams, whereas
potential density will be more nearly advected. The energetics corresponding to
that process is a final state in which the momentum field is smoothed, but the
potential density is inverted. This configuration gives the (hypothetical) final
mean state as $\phi = -Bz$, so double the final potential energy, and the more
difficult criterion $Ri < 1/4$.

We suppose that the first criterion is for the persistence of turbulence which
will ultimately mix the layer, while the more restrictive criterion is for the outbreak of turbulence in flow that was initially laminar. We do not imagine that
this actually results in the simple inverting of the temperature field but only that
this is an idealised end point to bear comparison with the complete elimination

of the shear. Perhaps the moral is that we should expect to see Richardson numbers generally large, but approaching 1/4, in layers of instability, where it will revert to a value of 1/2 as mixing proceeds.

In practice, the calculation of the shear and stratification is so dependent on the techniques used to analyse the data that such speculation is hardly capable of being tested. There is some suggestion in the data that the atmosphere might be stratified in distinct layers of constant Richardson number. It follows that when this data is sampled, using probes whose temperature and velocity response is different, there will be large fluctuations in the apparent Richardson number at the edges of the stratified layers, which seems to be what is found. Perhaps we should try to force the data to be piecewise linear in the same layers of shear and stratification, and see if that is more suggestive.

6.7 Some generalities on shearing instability

Using the energy argument, one might estimate the amplification rates in stably stratified shear flow. Equipartitioning the surplus energy deduced when potential density is on its way to being overturned between u'^2 and w'^2, gives the estimate

$$w' = h(\alpha^2 - 4g\beta)^{1/2}/6^{1/2} \tag{6.22}$$

but w'/h is the amplification rate for the displacement, so we estimate

$$\text{amplification rate} \simeq 0.4(\alpha^2 - 4g\beta)^{1/2} \tag{6.23}$$

From the Rayleigh problem we know the amplification for $\beta = 0$ so might even refine our estimate to

$$\text{amplification rate} \simeq 0.2\,(\alpha^2 - 4g\beta)^{1/2} \tag{6.24}$$

An extension of this type of theory is of use in the description of convectively unstable flow in the presence of shear as in Section 7.4.

6.8 Stratification outside the shear zone

If the stratification is confined to the layers outside the sheared zone and density is continuous at the interface, then only the process of propagation of energy by gravity waves is represented, and we expect the growth rates to be less because of this mechanism alone.

The Helmholtz problem can again be extended to suggest what might happen. We expect from the analysis of the previous chapter, that the stable layers will act to extend the action of the waves further from the seat of instability, which will surely decrease the rate of amplification.

The flow outside the sheared layer must now satisfy

$$\{(U - c)^2(\partial^2/\partial x^2 + \partial^2/\partial z^2) - gB\}\,\psi = 0$$

thus

$$\psi = A_1 \exp -\lambda_1 z \text{ for } z > h$$

and

$$A_2 \exp +\lambda_2 z \text{ for } z < h$$

where

$$\lambda^2 = k^2 - gB/(U - c)^2 \qquad\qquad (6.25)$$

and the λ whose real part is positive must be chosen, if this is possible. This is so, for example, when k is large. As in Section 6.1 continuity of pressure and displacement of the interface gives the frequency relation

$$(U_1 - c)^2\lambda_1 + (U_2 - c)^2\lambda_2 = 0 \qquad\qquad (6.26)$$

or

$$(U_1 - c)^2\left(1 - \frac{gB}{k^2(U_1 - c)^2}\right)^{1/2} + (U_2 - c)^2\left(1 - \frac{gB}{k^2(U_2 - c)^2}\right)^{1/2} = 0$$

$$(6.27)$$

Solutions are likely to be messy, but we can easily expand for k large, which gives

$$(U_1 - c)^2 + (U_2 - c)^2 - gB/k^2 + U^2 O(gB/k^2 U^2)^2 = 0 \qquad\qquad (6.28)$$

or

$$(c - (U_1 + U_2)/2)^2 = -((U_1 - U_2)/2)^2 + \frac{gB}{2k^2} + O(k^{-4}) \qquad\qquad (6.29)$$

This shows that short waves are damped by the stratification. It turns out that there is no term of order k^4 and the solution 6.29 is exact. However, as k becomes smaller the distinction between the real and imaginary parts of the square root become more debateable and we need to follow the solution with some care. Results of such a calculation are shown in Figure 6.5. It is convenient to take axes moving with the mean flow, which amounts to putting $U_1 = +U$, $U_2 = -U$, to use U as the unit for measuring c and $(gB)^{1/2}/U$ as the unit for measuring k, λ_1 and λ_2. We find that for k suitably large, c tends to i, λ_1 and λ_2 to k. By $k = 1$, $c = 0.71\,\mathrm{i}$, $\lambda_1 = 0.9 - 0.2\,\mathrm{i}$, $\lambda_2 = 0.9 + 0.2\,\mathrm{i}$ nearly. Not only is the motion penetrating deeper into the fluid as the wavelength increases but it is also acquiring a phase tilt. This is in the sense of giving a flux of energy away from the shear zone. Remember that the flow relative to the wave is in the opposite sense in the upper and lower layers which accounts for the slope of

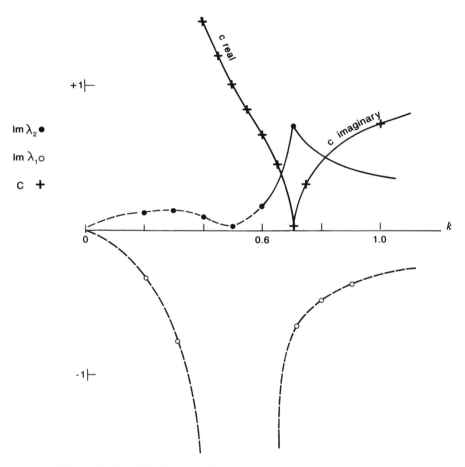

Figure 6.5 Stratification outside the shear zone serves to propagate the energy released by the shearing instability away from the shear zone as gravity waves. This gives the rather complicated picture of growth rate and propagation speed shown in the figure and discussed in the text in Section 6.8. When the stratification is in the shear zone but not outside, there is no propagation of energy but the vertical displacement of parcels is inhibited by the buoyancy, particularly for long waves. This is not a complicated picture.

phase being the same in order to account for energy propagation in opposite directions.

By $k = 0.8$, $c = 0.47\,\mathrm{i}$, $\lambda_1 = 0.61 - 0.51\,\mathrm{i}$, $\lambda_2 = 0.61 + 0.51\,\mathrm{i}$. At the critical value $k = (0.5)^{1/2}$ we arrive at $\lambda_1 = -0.71\,\mathrm{i}$, $\lambda_2 = +0.71\,\mathrm{i}$, $c = 0$. This is the end of the unstable regime and for longer waves c is real. For $k = 0.6$, $c = 0.62$, $\lambda_1 = -2.56$, $\lambda_2 = +0.15\,\mathrm{i}$. Thus we see a steering level near the upper layer, with phase tilt directing energy away from the shear zone on both sides, but more intensively in the upper layer than the lower. There is a corresponding solution for the lower layer with $c = -0.62$.

At $k = 1/2$, the steering level reaches the edge of the shear zone, and the wavelength in the upper layer goes to zero. For smaller k the waves relax again, so that by $k = 0.2$, $\lambda_1 = -0.37\,\mathrm{i}$, $\lambda_2 = 0.11\,\mathrm{i}$.

Waves of greater wavelength tend towards $\lambda_1 = -\mathrm{i}k$, $\lambda_2 = +\mathrm{i}k$. We might interpret this neutral regime as representing the steady conversion of shearing energy into wave motion which then propagates that energy to a very much deeper layer.

We wonder if the regime $(0.5)^{1/2} > k > 0.5$ might need more elaborate interpretation.

6.9 Shear with gravitationally unstable stratification

An interesting speculation concerns the interaction between dynamically unstable shear, and gravitationally unstable stratification. In many cases the growth of unstable perturbations converts either mean potential or mean kinetic energy into eddy energy, but not both. There is even some indication of this in the Helmholtz problem where it is not actually true, for unstable stratification does increase the amplification rate of the shearing instability. However the most rapidly amplifying motion is for large wavenumber where the mechanics is dominated by the velocity difference, no matter how small that is.

Similarly, when looking for a mechanism for creating fast-moving squall lines like those found in the tropics, we once thought that perhaps a very stable lower layer might support fast-moving gravity waves at its upper surface and that these might carry along fast-moving convection in an upper unstably stratified layer. Linear analysis shows that they cannot. Either we get stationary convection, or propagating gravity waves, but no mixture.

We might also speculate on the interaction between systems; what would happen if the radiating gravity waves were to meet a second system, a bit like the one that radiated them in the first place. Will they trigger off instability in the new system? Using the analysis of Section 6.8 as a guide, one suspects that this will happen only if the matching is rather precise. There is some, but incomplete evidence of this happening in real flows that are both stratified and sheared, with families of nearly unstable layers at about the same value of the air velocity.

While at first sight the application of proper physical boundary conditions, like no flow through impervious boundaries and no slip at rigid surfaces, seem obvious, this is not so in practice. Part of the problem concerns the variation in the scale of motion as a boundary is approached. A very simple example will make some aspects of that complexity clear.

6.10 Turbulence near the ground

For neutrally stratified flow over a rough surface, like the usual atmospheric conditions near the ground, flow is dominated by the transfer of momentum from the air above, towards the ground. The flow is usually turbulent and the mechanism for momentum transfer is through the action of eddies. At least in the absence of convection the eddies are mechanically driven. Close to the surface the vertical scale of the eddies is limited by the proximity of the ground. Thus if we invent an eddy diffusivity to represent the transfer process, we can suppose that the space scale of the eddies is comparable with their height above the ground, while the velocity scale is given by a characteristic speed u_*, where the stress τ is ρu_*^2. From dimensional arguments we might put the eddy diffusivity as $ku_* z$, where k is some non-dimensional number, which is going to be called Von-Karmans constant. The flux of momentum is then given by

$$\tau = \rho k u_* z \; \partial u/\partial z \tag{6.30}$$

Finally we see that there must be a layer, more or less close to the ground, in which the stress τ varies little, otherwise the flow would slow down catastrophically. This allows us to integrate giving

$$u = (u_*/k) \log (z/z_0) \tag{6.31}$$

and we identify z_0 as a constant defining the roughness of the surface. This profile is closely satisfied by observed winds in the gentle stratification that has been assumed. There is some doubt as to the zero of the reference level for the height when the roughness elements are tall; as with trees with tall trunks but dense canopies for example, whereupon a zero-plane displacement can be added to z.

While this seems a reasonable derivation, and it can be made more persuasive, it has several disturbing features. We might suppose the motion to be essentially two-dimensional in the plane defined by the mean flow and the vertical, as in Figure 6.6(a) and the turbulence to be generated as a result of mechanical instability of the shear layer. But if this were so, we would expect vorticity, rather than momentum to be transferred by the eddies. This would give

$$\partial \tau/\partial z = \rho k u_* z \; \partial^2 u/\partial z^2 \tag{6.32}$$

and imply that the mean velocity u varied linearly with height in the constant-stress layer. Observation shows that it does not. Philosophically this is valuable. It is all very well saying some law is a good fit, but much better to show that some other plausible relation is not so good.

Moreover the logarithmic velocity profile has $\partial^2 u/\partial z^2$ of the same sign everywhere, and Rayleigh's criterion shows that it would be stable to such 2-D perturbations.

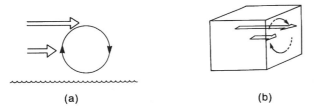

Figure 6.6 What simple realistic shearing instability probably looks like; not two-dimensional (carrying vorticity) as in (a) but essentially three-dimensional (carrying momentum) as in (b).

In order to transfer x-momentum rather than y-vorticity, the eddy motion must be mainly in the y–z plane, with the eddies carrying javelins of fluid between different heights as in Figure 6.6(b). The mechanism for such over-turning is not clearly related to the shearing instability of this chapter.

Again, if the eddy transfers momentum over the finite height range between height z and the ground, then the momentum gradient at z is not the most compelling way of representing the transfer. Rather it looks more like what is transferred is the velocity difference between height z and the ground, more like momentum of density $\rho(u(z_2) - u(z_1))$ transferred at speed u_*. If this is also a self-similar process we might expect a relation something like $z_2 = \alpha z$, $z_1 = z/\alpha$, where α is a similarity parameter, to give

$$\tau = \rho u_*^2 = \rho u_* (u(\alpha z) - u(z/\alpha)) \tag{6.33}$$

The solution of this finite difference equation with u_* constant is

$$u = (u_*/2 \log \alpha) \log(z/z_0) \tag{6.34}$$

and we recover the nice logarithmic solution that fits the observations. For the value of α that gives $2 \log \alpha = 0.4$ needed to make the fit good, we find $\alpha - 1/\alpha = 0.4$, which was the sort of shape we visualised for the eddies near the ground. We wonder what scale of roughness extracts the momentum. Experimental sites that give good logarithmic profiles are usually large expanses of uniform vegetation. It may be, at least over land, that it is only the larger roughness elements that matter. Ken Bignell and I once estimated that one tree per kilometer run of wind would extract as much momentum as the intervening grass. We imagined a tree of height h, with a substantial windspeed difference between upwind and downwind, thus a pressure difference of nearly $\frac{1}{2}\rho u^2$ con-tributing stress to be compared with that due to 'the grass'

$$\frac{1}{2}\rho u^2 h \simeq \rho u_*^2 L \tag{6.35}$$

where L is the distance to the next tree. With the typical value $u_* \simeq u/30$ we get $L \simeq 450\,h$ or maybe a little less to make up for incomplete stoppage of the wind by the tree, or one tree per kilometre.

Chapter 7

Vertical convection

7.1 Hydrostatics

We have in mind the processes of cumulus and cumulonimbus convection. They are characterised by having vertical and horizontal velocities of comparable magnitude, and pressure significantly non-hydrostatic. This is, at first sight astonishing. Air in cumulus convection attains vertical velocities of a few m s^{-1} at heights of several 100 m. The vertical acceleration is therefore about 10^{-2} m s^{-2} which is three orders of magnitude less than the gravitational acceleration g, yet this is not enough to allow the acceleration term in the vertical component of the momentum equation to be neglected.

Perhaps the simplest way of seeing why, is to recognise that we might want to eliminate the pressure by cross-differentiating the horizontal and vertical components of the momentum equation. This is the way we derive the vorticity equation, for example. In this application we use the vertical variation of pressure to eliminate the horizontal. When we differentiate the vertical component of the momentum equation

$$\frac{Dw}{Dt} + \frac{1}{\rho}\frac{\partial p}{\partial z} + g = 0 \tag{7.1}$$

with respect to x say, what was the 'large' term, that in g, vanishes identically, and so cannot be supposed large any more. The moral is that the quality of an approximation depends on what you want to do with it.

The same effect is found with surface gravity waves. If the pressure in the liquid were truly hydrostatic, then the amplitude of the wave would be independent of depth. But we know that the amplitude decreases with depth like

$\exp -kz$ where k is the horizontal wavenumber, independent of the smallness of the amplitude of the wave and therefore the smallness of the accelerations of the fluid.

Another approximation occurs when the horizontal scale is very small compared with the vertical scale. Under these circumstances $\partial p/\partial z$ is determined by the 'environment' density ρ, while ρ' is the parcel density, and we recover the Archimedean form of equation 7.1:

$$Dw/Dt \simeq g(\rho' - \rho)/\rho \tag{7.2}$$

7.2 Effect of molecular diffusivity

Consider an incompressible fluid, defined as one in which particles retain their density, but which density may vary from parcel to parcel. Suppose that the motion is confined to the x–z plane, neglecting Coriolis accelerations, but retaining molecular thermal diffusivity k_m and viscosity ν_m for the moment. The Boussinesq equations become

$$D\mathbf{v}/Dt + \nabla(p'/\rho) - g\,\mathbf{k}\,\phi' = \nu_m \nabla^2 \mathbf{v}$$
$$D(\phi + \phi')/Dt = k_m \nabla^2(\phi + \phi')$$
$$\partial u/\partial x + \partial w/\partial z = 0 \tag{7.3}$$

whence by cross-differentiating the momentum equation and using a stream function ψ for which $u = \partial\psi/\partial z$, $w = -\partial\psi/\partial x$, we get the vorticity and thermodynamic equations

$$\frac{D}{Dt}\nabla^2\psi + g\frac{\partial\phi'}{\partial x} = \nu_m\nabla^4\psi$$
$$\frac{\partial\phi'}{\partial t} - bw = k_m\nabla^2\phi' \tag{7.4}$$

as in equation 5.4 except with diffusive terms added. We wonder what would happen if a basic state, $\partial\phi/\partial z = -b$, in which potential temperature decreases uniformly with height everywhere, was perturbed. Linearising equation 7.4, we get

$$\frac{\partial}{\partial t}\nabla^2\psi - g\frac{\partial\phi'}{\partial x} = \nu_m\nabla^4\psi$$
$$\frac{\partial\phi'}{\partial t} - bw = k_m\nabla^2\phi' \tag{7.5}$$

Taking the separable, constant-shape solutions

$$\psi = A\,e^{\sigma t}\,\exp ikx\,\sin \pi z/H$$
$$\phi' = B\,e^{\sigma t}\,\exp ikx\,\cos \pi z/H \tag{7.6}$$

gives

$$(\sigma + v_m m^2) m^2 A + ikB = 0$$
$$(\sigma + k_m m^2) m^2 B - ik\,gbA = 0$$

where

$$m^2 = k^2 + \pi^2/H^2, \tag{7.7}$$

or

$$(\sigma + v_m m^2)(\sigma + k_m m^2) - k^2\,gb = 0 \tag{7.8}$$

The non-diffusive solution, with $v_m = k_m = 0$ is

$$\sigma^2 = gbk^2/(k^2 + \pi^2/H^2) \tag{7.9}$$

Apart from the sign of b and σ^2 this is the same as the solution for stable gravity waves containing the same terms and therefore, the same basic mechanics. The contrast between stable and unstable equilibrium is like that of a rigid rod having a small disturbance about its equilibrium position. The relevant equation contains the same physical parameters, whether the rod is suspended from the bottom or from the top. But suspended about its lower end the rod is in an unstable equilibrium and the development of any displacement is exponential in time whereas, suspended about its upper end it is stable, and the displacements are sinusoidal in time. The term k^2 in the denominator disappears when the hydrostatic approximation is made whether the flow is stable or unstable. Maybe it is astonishing at first sight that the flow amplifies faster with the hydrostatic approximation. But notice that the horizontal gradient of pressure is then not used to accelerate the air, so all the energy available is used for vertical acceleration. For cumulus convection, temperature excesses of about 0.2 K are typical over a height of several hundred metres, so the amplification rates are of order $1/200$ s which is not unrealistic. Maximum amplification rate occurs at a scale for which $k \ll \pi/H$, which implies columns with separation distance much less than H, whereas cumulus tend to have separation distance of about $3H$ so this aspect is unrealistic.

As Figure 7.1 illustrates, dissipation prevents the columns becoming indefinitely narrow for, as equation 7.8 shows, as k becomes large the dissipation terms dominate the equation for σ. From the form of this equation we can see that maximum amplification rate will then be for k nearly equal to $(gb/v_m k_m)^{1/4}$. The figures used above give k about $30\,\text{m}^{-1}$ or an updraught 10 cm wide, rather than infinitely narrow. This is still quite absurd as a description of real cumulus.

Estimates of the Rayleigh number appropriate for the free atmosphere vary greatly. Estimates of diffusivity alone range from molecular diffusivity of $v \simeq 3 \times 10^{-5}\,\text{m}^2\,\text{s}^{-1}$ and eddy diffusion of $10\,\text{m}^2\,\text{s}^{-1}$. Values of H vary from a few centimetres for the ground boundary layer to 100 m for a convective 'bubble' to 1000 m for a whole cumulus cloud.

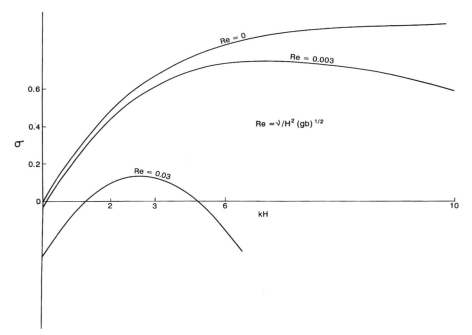

Figure 7.1 Simple convective instability. For Re = 0, motion of short
wavelength amplifies at an Archimedian growth rate independent of
wavelength. Motion of long wavelength is inhibited because some of the
potential energy released by vertical displacement must be used to carry air
large horizontal distances. Diffusion of temperature and of momentum
decreases the growth rate, particularly for motion of smaller scale.

One school of thought has it that we should replace the molecular diffusiv-
ities by values typical of turbulent exchange such as $10 \, \text{m}^2 \, \text{s}^{-1}$, which gives
$\pi/k \simeq 100 \, \text{m}$. While now being more realistic, at least for the description of
the individual bubbles of warm air, of which we might imagine cumulus as being
composed, this theory seems largely to avoid the issue of size determination.
After all it is the convection itself that determines the intensity and scale of the
turbulence; it is not defined beforehand like molecular diffusion is. Perhaps the
best we can do is to say that vertical velocities of about $1 \, \text{m} \, \text{s}^{-1}$ are generated in
a typical cumulus, in which the updraughts are 100 m across, *because* of the
value of k_{eddy} and v_{eddy}, which are given by an expression of the form

$$k_{eddy} \simeq v_{eddy} \simeq 1 \, \text{m} \, \text{s}^{-1} * 100 \, \text{m} * efficiency \tag{7.10}$$

This is a cunning way of introducing some non-linear effects into a linear
treatment, and such development leads to interesting speculations, at present
unexplored. The steady eddy regime on long jets (discussed in Section 6.4) is one
example.

7.3 Really-shallow convection

When H is very small, molecular effects are properly significant. The biologist Benard, peering down his microscope at a strongly illuminated slide, discovered patterns which he thought at first were life-forms, but which turned out to be shallow convection cells. These have been described in terms of the steady solutions which have $\sigma = 0$.

The theory of the previous section is not very good because it makes no reference to boundary conditions, and it should. It turns out to be acceptable for the rather peculiar conditions of rigid, perfectly slippery, isothermal lids ($w = \partial w/\partial z = \phi' = 0$ at $z = 0$ and H), but it will do to show the principles involved. Suppose we put $\sigma = 0$, and then examine the resulting equation

$$k_m v_m (k^2 + \pi^2/H^2)^3 - k^2 gb = 0 \tag{7.11}$$

to find the minimum value of gb for variation in k. We then identify this with a critical stratification for the outbreak of steady convection. We discover the non-dimensional number

$$R_a = gbH^4/v_m k_m \tag{7.12}$$

named after Rayleigh who identified its importance. For the problem posed here it has the minimum value $27\pi^4/4$. Notice that it is one of those non-dimensional numbers that mathematicins know of as 'of order unity' even when, as here for example, it has the numerical value 658. This goes with the value $k \simeq \pi/\sqrt{2}H$, which gives a realistic aspect ratio for the cells of about $3 : 1$. We notice in passing that this ratio is independent of the value of the diffusivities.

7.4 Convection in shear

Tall narrow columns are likely to be disrupted by shear, so perhaps that is why we do not see them. With a mean wind, $U = az$ along the x-axis, the equation without molecular diffusion reduces to

$$\partial^2 \psi/\partial Z^2 - (1 - r/Z^2)\psi = 0 \tag{7.13}$$

where $Z = (kz - \sigma/a)$. Fundamental parameters are $r = gb/a^2$ which is a (negative) Richardson number, and kH, the non-dimensional wavenumber. The general solution of equation 7.13 is a Whittaker function. These are in general, rather complicated, especially in the complex plane where the amplifying solutions take us. However solutions can be expressed in finite terms when $r = n(n+1)$ and n is an integer. The value $n = 1$ is the first solution both tractable and interesting. This gives

$$\psi = Ae^{Z}(1 - 1/Z) + Be^{-Z}(1 + 1/Z) \tag{7.14}$$

and the boundary condition of rigid lids demands

$$\sigma^2 = a^2(1 + k^2H^2/4 - kH \coth kH) . \tag{7.15}$$

Curiously, this solution, shown in Figure 7.1 is the same equation as for the classical Eady problem of baroclinic stability and like it, shows that short waves are stable. In realistic cumulus situations r is not very well defined, and is probably quite large, which suggests that shear may not be the main mechanism curtailing the short wavelength instability. But notice that we do not need persistent large-scale shear, but only near our developing convection, at the time it is developing. Thus the cells might be disrupted by some general background of fluctuating shear; a possible stochastic influence. We conclude that environmental shear is not a particularly attractive mechanism for the determination of the scale of cumulus convection. If it were then there would be the possibility of interesting feedback between the basic shear and the convection acting in the sense of changing it, either by mixing it away, or by transferring momentum against the mean gradient and unmixing it.

7.5 A possible linear selection process

Partly on account of its novelty and character of being dispersive in the growth rate rather than in the phase speed, we suggest the following mechanism for the selection of the horizontal scale of cumulus. A local initial disturbance of unsheared statically unstable flow can be written as

$$\psi = \int f(k) e^{\sigma t} e^{ikx} \sin\left(\frac{\pi z}{H}\right) dk \tag{7.16}$$

where $f(k)$ is a slowly varying function of k, and σ is given by equation 7.9 which shows that all waves except the very long ones grow at the same rate. Thus, apart from this deficiency in the long waves, the initial disturbance increases in amplitude but keeping the initial shape unchanged. To see what this implies, consider an initial state

$$\psi = \frac{a \sin(\pi z/H)}{(a^2 + x^2)} \tag{7.17}$$

so that $f(k) = \exp -ak$, where a is not large, and let us represent σ as a two-part approximation;

$$\sigma = kn_0/k_0 \text{ for } 0 < k < k_0$$
$$= n_0 \quad \text{for } k_0 < k$$

where $k_0 = \pi/H$ and $n_0 = (gb)^{1/2}$. By this device we are able to write down the solution in finite terms as the real part of

$$\psi = \left(\frac{\exp{(n_0 t - k_0 a + ik_0 x)}}{a - ix} + \frac{1 - \exp{(n_0 t - k_0 a + ik_0 x)}}{a - ix - n_0 t/k_0} \right) \sin k_0 z$$

(7.18)

For times such that $n_0 t \gg a k_0$, the first term dominates, and this term shows propagation of a wavelike disturbance, with wavenumber k_0 through the width of the system. For this time, $\sigma t \gg a k_0 = \pi a/H$, is likely to be large, so the initial local disturbance is likely to be well into its life cycle; we expect to see an extinct cumulus at the centre of a rather better organised pattern of new cumulus with dominant space scale k_0.

One gets the superficial impression of such propagative development away from an ancient origin on some occasions of real convection, and the proposed mechanism has some attractive features like not requiring more parameters. It is salutory to see that a spectrum with some frequencies missing looks like a sample of just those frequencies, at least in some regions of space.

7.6 Steady convective overturning

The difficulty of collecting detailed observations of intense travelling cumulo-nimbus suggested to Ludlam that these storms might be thought of as steady in the sense of propagating relatively unchanged through the network of observing stations. This way he could replace each individual station by the time section extended in space at the propagation velocity. His analysis of his extensive observations and experience suggested a flow field, relative to the storm, like that shown in Figure 1.6. Moist potentially warm air rises over dry potentially cold air. Rain falls from the warm updraught, releasing it from the weight of liquid water so enhancing its buoyancy. This rain falls into a downdraught which is cooled by evaporation and loaded by the weight of the liquid water to give it downward directed buoyancy.

A 2-D, incompressible, dry adiabatic model brings out some interesting features. Choosing axes moving with the storm, we have $(\partial/\partial t)_{rel} = 0$, by definition, and streamlines coincide with trajectories. Because a stream function ψ exists for two-dimensional motion, the thermodynamic equation can be integrated directly

$$D\phi/Dt = 0 \quad \text{so} \quad \phi = \phi(\psi)$$

(7.19)

Similarly for the 2-D vorticity equation

$$\frac{DN}{Dt} = -g \frac{\partial \phi}{\partial x} = -g \frac{d\phi}{d\psi} \frac{\partial \psi}{\partial x}$$

so

$$N = F(\psi) + gz \frac{d\phi}{d\psi} \tag{7.20}$$

where F is a function as yet undetermined. Now suppose that the inflow to the storm has shear a and a profile of potential temperature such that $\partial\phi/\partial z = -b$, which is to represent the excess potential temperature that would be released by the moist buoyant ascent in a realistic system. This information defines the propagation speed of the storm. We choose to represent this in terms of the height of the steering level $z = z_*$ where by definition, the flow relative to the storm vanishes. Far away on the upstream side, in the slab defined by $0 < z < z_*$, we have, by definition

$$\psi = \frac{1}{2} a (z - z_*)^2 \text{ and } \phi = -bz \tag{7.21}$$

It follows from equations 7.19 and 7.20 that

$$\phi = -bz_* + b(2\psi/a)^{1/2} \text{ and } F(\psi) = a + gb/a - gbz_*/(2a\psi)^{1/2} \tag{7.22}$$

The strategy is to follow a parcel into the storm up and out above the inflow, carrying the vorticity and potential temperature relations 7.22 with it. The outflow must satisfy the vorticity equation, but remote from the system the term in $\partial^2\psi/\partial x^2$ might be neglected compared with $\partial^2\psi/\partial z^2$, so ψ must satisfy the ordinary but non-linear differential equation

$$\frac{\partial^2\psi}{\partial z^2} = a + \frac{gb}{a} + \frac{gb(z - z_*)}{(2a\psi)^{1/2}} \tag{7.23}$$

The solution must satisfy boundary conditions

$$\psi = \partial\psi/\partial z = 0 \text{ at } z = z_*$$
$$\psi(z = 0) = \psi(z = H) \tag{7.24}$$

It can be verified that the required solution is

$$\psi = \frac{1}{2} a \beta^2 (z - z_*)^2$$

where

$$\beta(1 - \beta) + gb/a^2 = 0 \text{ and } z_* = \beta H/(1 + \beta) \tag{7.25}$$

We are astonished that the ugly non-linear equation 7.23 has such a simple solution. I discovered it when my finite difference approximations to a numerical solution using successively finer grids, gave the same answer. This could only be true if the truncation errors were identically zero, so the solution had to be a polynomial of low order.

We notice the reappearance of the Richardson number; $r = gb/a^2$, and that we must have $r > -1/4$ for the solution to exist, consistent with the criterion for the onset of shearing instability in face of stable stratification.

For $r = 0$, we have $\beta = 1$, and $z_* = H/2$; the neutrally stratified flow is turned round, but is otherwise unchanged. For $r > 0$ the inflow is convectively buoyant $\beta > 1$, and $z_* > H/2$. Thus the steering level is above the middle level, with the outflow moving faster, with more shear than the inflow. One interpretation is that the horizontal gradient of temperature inside the storm has added vorticity to the air during its passage through the system, and another interpretation is that the air has been accelerated through the action of buoyancy and of pressure forces.

On the downdraught side, $z_* < H/2$, and the upper-level inflow is accelerated into a more sheared outflow at the ground. We notice that the inflow at both upper and lower levels has been increased, compared with that before the storm. Technically it is the whole flow to infinite distance which has been accelerated. It is a disquieting feature of every steady system that they have had time enough to do energetically outrageous things like this. Initial value versions of this problem show that the increased inflow propagates outward at the speed of some appropriate gravity wave. Our assumption that $|\, \partial^2 \psi / \partial x^2 \,| \ll |\, \partial^2 \psi / \partial z^2 \,|$ demands that small-scale gravity waves are not energetically significant.

We have chosen to take the vorticity of the ambient flow as given because the vorticity of the large-scale flow is modified relatively slowly. How it is modified is the subject of mesoscale dynamics.

An alternative derivation of equation 7.25, notes that the Boussinesq form of the energy equation is also valid along the streamline going into and out of the system. This contains an important term accounting for the work done by the pressure, but remote from the system we can calculate the pressure difference between the two streamlines using only the hydrostatic equation. Continuity of mass then closes the problem. This analysis extends rather readily to elaborate temperature profiles, more realistic thermodynamics, and variation in static density. It might even make an impression on the 3-D problem.

7.7 Updraught slope

While this theory has many attractive features, one germinal point emerges when we try to solve equation 7.20 to find the dividing streamline, and other details of the flow near the centre of the system. We find that the dividing streamline slopes the opposite way to that expected. As shown in Figure 1.6(b), the warmer air now lies below the cooler. Also, the water now cannot fall out of the updraught into the downdraught to cool it and accelerate it down. Having obtained this result one can readily see it must be so. The original configuration demands that that slowly moving air near the ground be taken round a gentle bend, accelerated substantially through buoyancy, then taken round a sharp bend at the top of the storm. Thus the air goes round shallow bends slowly, sharp bends quickly. Also notice that at this same top corner we find the slow-moving, potentially cold, upper level inflow is taken round a gentle corner by the same pressure field that takes the fast inflow air round a sharp bend. This argument can be dignified by using the Bernoulli equation along each of the dividing streamlines, noting that adjacent parcels share the same pressure field.

Reversing the slope of the interface removes this qualitative objection by making the gently moving air go round sharp corners, fast air round gentle ones.

A number of hypotheses have been advanced to account for the discrepency. One is that the warm inflow is potentially colder than the environment below cloud base, so decelerates, and is able to lie over the cold air, which is locally warmer. Alternatively, it may be that the intermittency which is characteristic of real systems is essential, with updraughts dropping their load of water into potential downdraughts, which then scoop up low-level air into new updraughts. Such a picture removes some of the constraints of the steady theory, but at the expense of much greater complexity.

I prefer the idea that the system is essentially steady, but vitally three-dimensional. If the updraught slopes in the same sense as the shear it can overlie a downdraught only if we twist the two in the third dimension, as shown in Figure 1.6(c). The original conditions determining the flow, generalised suitably to include the third dimension, need to be supplemented by another relation, which might be the conservation of potential vorticity. Indeed one could envisage the 3-D steady overturning problem as being well posed by such a set. One nice feature is that such a system clearly has parity; there would be left- and right- handed storms. Such are observed in the middle-west USA, with tornado development likely in one but not the other.

7.8 Real convection

Above a smooth hot surface, we can see a superadiabatic region, where heat transfer is through the agency of molecular conduction because advection is hindered by the effect of molecular viscosity and the nearness of the solid surface, as represented by the Rayleigh-type of description, though radiative transfer might also be important.

Above a realistic ill-defined, bumpy, re-entrant hairy, surface there is likely to be a much more confused assembly of physical processes whose result is to transfer momentum input to:

- a logarithmic layer, where transfer is by motion whose form is conditioned by the closeness of the boundary, and which is hardly aware of buoyancy forces, and then to

- a layer of 'freer' convection where the motion begins to become aware of the buoyancy, but still recognises the lower boundary (associated with the names of Monin and Obukhov) then on to
- the layer up to cumulus base where buoyancy probably dominates but mixing of ascending and environmental air is important, on to
- the top of the cumulus layer, where condensation into cloud particles takes place, shear becomes important, and mixing is so complete as to reduce the buoyancy practically to zero and give the 'synoptic' rule that cumulus top will be where the environment temperature has the same slope as the wet adiabatic;
- after the action of 100s of cumulus, or a couple of days of convective activity, something of larger scale empties the cumulus convective boundary layer. This might be a cumulonimbus, steadily peeling off the moist air, and creating kinetic energy on a fairly small vertical scale. It might be on the scale of fronts where statically stable lifting injects the proceeds of latent energy into the large-scale quasi-geostrophic flow.

Chapter 8

Mesoscale motion

8.1 Definition of mesoscale

There is no universally accepted definition of mesoscale. Here we use it to denote motion that is not dominated by the mechanics of gravity waves, nor by the near geostrophy of Chapter 9, but is aware of, and compromises between, both. We have in mind the motion in frontal zones, the organisation of several cumulonimbus to make a severe storm, squall line, or even a hurricane, and the longer-term evolution of sea breezes. Sometimes the word mesoscale is used to indicate motion whose scale is of order 100 km but this geometrical definition includes a number of distinctly different physical processes. For example, motion on the scale of 100 km in the ocean is very similar in mechanism to the scale of 1000 km in the troposphere in that baroclinity, static stability, and variation of Coriolis parameter with latitude play similar roles. We prefer to use a criterion based on the dominant physical processes rather than on the geometry. There are disadvantages, and we find that the meridional overturning, of global scale, comes within our new definition of mesoscale, which is at first sight astonishing. There is often no very obvious energy source for mesoscale motion defined this way. Rather the effect is to organise the energy already available. Thus examples of mesoscale motion would include the organisation of several cumulonimbus clouds, perhaps through the interaction of their downdraughts, or diurnal land–sea temperature differences modulated to produce mean flows of longer timescale, or the effect of the deformation produced by flow of much larger scale. A classic example of such distortion is where air at upper levels flows into a jet. The geostrophic,

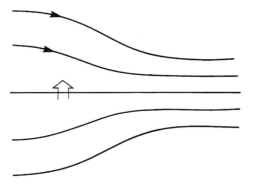

Figure 8.1 x–y section showing air entering a jet must cross isobars in order to change speed; at the jet entrance, the air accelerates so must cross the isobars towards low pressure. Flow will again approach geostrophic further down the jet. Geostrophic flow is nearly non-divergent and this comparatively weak ageostrophic flow is very important for generating convergence/divergence.

essentially non-divergent, flow accelerates as it enters the jet as in Figure 8.1, but in order for the parcel of air to accelerate there must be ageostrophic flow allowing the air to cross the isobars towards low pressure. This ageostrophic, divergent/convergent component of the flow leads to descent on the cold side of the jet entrance and ascent on the warm side; a direct circulation in the sense that the potentially colder air descends while potentially warmer ascends, consistent with a decrease in available potential energy and the observed increase in the kinetic energy of the jetstream air. This adjustment flow is what we will refer to as the mesoscale part. Again, Bergeron suggested that a confluent wind field, by bringing isotherms close together, might be the mechanism for forming fronts. Perhaps the simplest mesoscale motion concerns the interaction between the quasi-geostrophic flow outside the frictional boundary layer, and the motion in the boundary layer itself.

8.2 Above the logarithmic layer

Observed wind profiles in nearly convectively neutral conditions suggest that the logarithmic profile is a reasonable fit only over the first 100 m or so above the ground. We wonder if this layer might provide a useful boundary condition for the flow above. Above the layer of constant stress, the effects of rotation and pressure gradients become significant. If the magnitude of the substantial acceleration $|D\mathbf{v}/Dt|$ is small compared with the Coriolis acceleration $|f\mathbf{v}|$ in the free atmosphere, then, since it is quadratic in velocity, we expect it to be small in the lower layer too and put

$$\rho f\, \mathbf{k}{\wedge}\mathbf{v} + \nabla p = \partial \tau/\partial z \tag{8.1}$$

where τ is the horizontal vector stress. The stress has usually become small compared with that at the surface, at a height of a few hundred metres. Thus if the stress decreased uniformly with height, it would change by some 20 to 30 % say in the first 100 m, and the assumption of constant stress for the purpose of integrating the equation for the profile of mean velocity, is nicely satisfied. It is a very different matter if we want to *differentiate* the stress as in equation 8.1 to find something else. In fact the stress usually varies more rapidly as we get nearer the surface so as one of my students pointed out, the constant stress layer is that layer where the stress varies most rapidly.

8.3 The Taylor-Ekman layer

Equation 8.1 can be solved to find the velocity, given the pressure field, so long as we specify a relation between the stress and the mean wind. Taylor and Ekman each supposed by analogy with molecular diffusion, that

$$\tau = -K \, \partial \mathbf{v} / \partial z \tag{8.2}$$

with K constant. Substituting this in equation 8.1 gives a pair of simultaneous scalar equations in $\mathbf{v}$, each of second order in z. Solution therefore demands two vector boundary conditions. Possible solutions include ones increasing exponentially with height. So if we are dealing, realistically, with a comparatively deep layer, going up in the atmosphere, down in the ocean, one condition is simply that $\mathbf{v}$ should remain bounded with increasing distance from the ground. This is a vector condition so counts as two scalar conditions.

The other boundary condition concerns behaviour at the rough boundary. Ekman supposed by further analogy with the molecular situation, that the tangential component of velocity should vanish at the ground. A complete explicit solution can then be found. I have always had misgivings about the plausibility of the no slip boundary condition for molecules near rigid boundaries. I do not doubt the experimental evidence as provided by Poiselles experiment, but choose to regard it as still astonishing. For the present situation, Taylor pointed out that we cannot use equation 8.2 all the way down to ground with K constant, because the eddy diffusivity must tend to zero as the flat but rough, ground is approached. For instance we need to be able to put a logarithmic boundary layer between this level and the ground. He proposed that the boundary condition should be visualised as applied at the top of the constant stress layer, where the stress should be continuous. To make this a sufficient condition we have to specify a stress law, which relates the stress to some property of the mean flow. The Newtonian form $\tau = \rho C_D \, | \mathbf{v} | \, \mathbf{v}$ might be used, or the linear approximation $\tau = \rho C' \mathbf{v}$. Not surprisingly Taylor's solution is a generalisation of Ekmans, to which it tends as the surface friction coefficient increases. An interesting feature of both of these theories is that the flow in the friction layer is across the isobars towards low pressure. Thus a parcel of air, deprived by friction of the speed that served to balance the pressure gradient by the Coriolis force, must accelerate towards lower pressure. With $\tau = 0.1\,\mathrm{N\,m^{-2}}$, we deduce cross-isobar winds of about $1\,\mathrm{m\,s^{-1}}$ in a layer 1 km deep which is about right. If we calculate from this the mass flux across the isobars we deduce a timescale of about 1 day to fill up a depression of radius 1000 km and a central pressure 10 mb less than the surroundings. This is much too violent, and what we have done in this calculation is to remove the pressure field but leave the velocity field unchanged; we cannot do this in a geostrophically balanced system, and the mesoscale motion intervenes.

8.4 Communication with the free atmosphere

Let us do it more properly. The layer in which the stress decreases with height and the free atmosphere above, communicate through the vertical velocity at the common boundary. Taking the curl of equation 8.1 eliminates the pressure field and converts the term containing the Coriolis acceleration into a divergence.

$$\frac{\partial}{\partial z}(\mathbf{k}.\mathrm{curl}\tau) = \mathbf{k}.\mathrm{curl}(f\mathbf{k} \wedge \rho\mathbf{v}) = \frac{\partial}{\partial x}(\rho u) + \frac{\partial}{\partial y}(\rho v) = -f\frac{\partial}{\partial z}(\rho w) \qquad (8.3)$$

where the time derivative of density in the equation of continuity of mass is negligible and the variation of f with the north–south coordinate has been ignored. Equation 8.3 can be integrated w.r.t. z from the ground where $w = 0$, to the top of the frictional boundary layer, where τ is small compared with that at the ground. This gives the vertical velocity at the top of the friction layer and therefore at the bottom of the free atmosphere, w_{top} in terms of the surface stress τ_s as

$$w_{top} = \frac{1}{\rho f}\mathrm{curl}_z\tau_s \qquad (8.4)$$

Taking the simple linear stress law $\tau_s = \rho v_* \mathbf{v}$ with v_* constant gives, as the boundary condition at the bottom of the free atmosphere

$$w_{top} = v_* \zeta / f \qquad (8.5)$$

where ζ is the vertical component of the vorticity. Suppose that the flow is bounded by a smooth rigid lid at $z = H$, to represent the constraining effect of the stratosphere. In the vorticity equation we replace the term $\partial w / \partial z$, by $-w_{top}/H$ and the vorticity equation becomes

$$D\zeta/Dt = -v_* \zeta / H \qquad (8.6)$$

which has solutions proportional to $\exp -v_* t / H$ with a damping timescale H/v_* of about a week with v_* of a few $\mathrm{cm\,s^{-1}}$. While this solution needs some qualification, for the change of phase and amplitude of ζ with height has been plausibly, but not rigorously neglected, it shows semi-quantitatively how the vertical velocity induced at the top of a shallow frictional boundary layer damps out flow throughout the whole quasi-geostrophic layer, with a realistic time-scale. We notice the suggestion that the timescale of the damping is independent of horizontal scale, except perhaps where the horizontal scale is so small that vertical scale H depends on it. In practice we suppose that this frictionally driven part of the stress can be added to the orographic contribution, and used as the lower condition for tropospheric flow.

8.4.1 Sverdrup flow

If we integrate equation 8.3 from the lower to the upper limit of the system, at both of which there is no vertical flux of mass, we get a condition on the surface stress $\mathrm{curl}_z \tau = 0$; essentially that there can be none. Estimating the size of terms assures us that the result is not pathalogical. That there should be a constraint on the surface stress seems an astonishing result to me. In Section 15.1 we identify the eddy transport of momentum $\overline{u'v'}$ as an important missing term in the atmosphere, and deduce that convergence of eddy transfer of momentum (the Reynolds stresses) is needed to drive the surface winds against friction, and deduce that there would be no surface winds if these stresses were absent.

Oceanic eddies are much less powerful than atmospheric eddies, and we can identify another term that is of interest there. This comes from the variation of Coriolis parameter with latitude. Including this term we get

$$\left(\frac{\partial \tau_y}{\partial x} - \frac{\partial \tau_x}{\partial y} \right) = \beta \int_0^H \rho\, v\, \mathrm{d}z \tag{8.7}$$

Sverdrup argued that the wind stress was balanced by this term at least in the eastern part of the ocean. This results in ocean currents nearly following the bands of easterly and westerly winds, and seems a very complicated way of arriving at this not-unreasonable result. An interesting feature is that there *is* now a mass flux in the west to east direction. This arrives at the western side of the oceans at about the right latitude to feed the western boundary currents, like the Gulf Stream and the Kuro-Shiwo, so these are available to carry the mass flux back to the eastern ocean. But they also have to do something with the vorticity that has been injected into the water by the wind stress over the eastern part. In the steady state, this must be disposed of ultimately by friction, possibly at the bottom, possibly at the shore, possibly by horizontal transfer to other ocean water, likely all three. If we have a frictionless ocean, driven from rest by a windstress curl, then the western boundary current becomes inexorably nar-rower and faster and in this case it is likely that instability and eddy transfer of momentum is ultimately important, concentrated in the western part of the ocean by the Sverdrup flow. Transfer of momentum between Gyres is also likely.

If we also integrate across the system with respect to x as we do in Chapter 15, then this term must vanish because it is proportional to the total mass flux across the x-direction, and this makes some statement about the angular momentum of the system.

8.5 Two-dimensional analysis

To a first approximation some consequences of mesoscale forcing can be treated as two dimensional. In Chapter 5 we noted that, for motion of short timescale which varied in only two space directions (x, z) say, the motion in that plane was independent of the y-component of velocity. This is no longer true when the timescale becomes so long that Coriolis accelerations are significant. Indeed this is the feature that we have singled out to be the criterion for our definition of the mesoscale. We assume that the vertical to horizontal aspect ratio of the motion is sufficiently large that the hydrostatic approximation to the pressure is adequate. The equations of Chapter 5 with Coriolis terms added and allowing for the possibility of a source of momentum in the y-direction (F), and diabatic heating Q are

$$\frac{Du}{Dt} + \frac{\partial}{\partial x}\left(\frac{\delta p}{\rho}\right) - fv = 0 \tag{8.8}$$

$$\frac{Dv}{Dt} + \frac{\partial}{\partial y}\left(\frac{\delta p}{\rho}\right) + fu = F \tag{8.9}$$

$$\frac{\partial}{\partial z}\left(\frac{\delta p}{\rho}\right) - g\phi = 0 \tag{8.10}$$

$$D\phi/Dt = Q \tag{8.11}$$

and

$$\text{div } \mathbf{v} = 0 \tag{8.12}$$

where the effect of the variation of mean density with height has been neglected.

8.6 The cold slab

Let us start with a simple analysis of a familiar concept. Consider a block of air near the ground which starts off with a slightly larger density than the surroundings. This might be due to cooling by the evaporation of precipitation from a group of convective storms, or by different heating over land and sea soon after sunrise, for example. Initially we choose to neglect momentum sources and further heating so put $F = Q = 0$. We suppose that the slab is initially of depth h_0 and of log-potential temperature anomaly in the horizontal $\delta\phi$. Conservation of energy and a little imagination, suggests something about its evolution. It would seem reasonable to suppose that the slab would tend to spread out roughly as in Figure 8.2. Let this spreading be characterised by a horizontal velocity u_0 at the ground varying with height to be reversed at height h_0. We might put

$$u = u_0 \frac{(h_0 - 2z)}{h_0} \quad h = h_0 \frac{(L - x)}{2L} \quad \text{for } -L < x < L$$

with $u = 0$, h constant elsewhere, and where $dL/dt = u_0$ (8.13)

Notice that L is a parameter of our picture of the motion and as such, is a function of time only, which justifies use of the ordinary derivative. The evolution of the tangential component of velocity is a little more difficult. Equation 8.9, with $\partial(\delta p/\rho)/\partial y = 0$ can be integrated to give

$$v + fx = \text{constant} \tag{8.14}$$

for each particle. But we notice that the surface air at $x = -L$ has just not yet moved, whereas that now at $x = L$ came from $x = 0$ at $t = 0$. Thus we have $v \simeq (L + x)f/2L$ at $z = 0$, varying linearly with height to have the opposite sense at $z = h_0$. These expressions allow us to estimate the terms in the equation for conservation of mechanical energy in the absence of dissipation

$$\int \left(\frac{1}{2}(u^2 + v^2) + g z \,\delta\phi \right) dx\,dz = \text{constant}$$

giving

$$u_0^2 h_0 L/6 + h_0 f^2 L^3/6 = g\,\delta\phi h_0^2 L/3 \tag{8.15}$$

Notice that using more complicated profiles of velocity and heating will merely change the values of the numerical coefficients in this equation. Using equation 8.13 reduces this to the non-dimensionalised form

$$dL/dt = (L_0^2 - L^2)/t_* \tag{8.16}$$

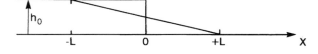

Figure 8.2 Schematic adjustment of a cold slab. In Section 8.6 we explore the consequences of supposing that the potential energy released as the cold air sinks relative to the warm air is organised this way.

which can be integrated analytically. We notice that while L is small $u_0 \simeq (2g\,\delta\phi\,h_0)^{1/2}$, which with multipliers comparable with 2 is consistent with observation and more elaborate theories for this 'dam break' problem. As L increases towards the value $L_0 = (2g\,\delta\phi\,h_0)^{1/2}$, u_0 tends to zero, so this is the space scale we identify with our definition of mesoscale. It is of the right size to describe the cloudless band extending over both land and sea near coasts often seen on satellite pictures. Solutions to equation 8.16 are sinusoidal in time and demand that the slab retrace its steps after the extreme displacement to return eventually to its original form. The driving force for this behaviour is that, at the extreme displacement, it has overstepped the geostrophic wind balance by a factor of two; another example of the rigid pendulum of Section 7.2. We imagine that some (but maybe uncritical) dissipation slows the advance before this happens.

The effort of introducing friction is illuminating. There will be frictional interaction with the ground, also interaction between the air in the slab and the air above. We have supposed that the flow speeds are similar at the top and bottom of the slab, but the penetration of slab air into upper air is likely to be much easier than penetration into the ground. Further observational and theoretical work indeed identifies the turbulent regime at the top front of the advancing wedge as being important. We might begin to extend the analysis by recognising that the total dissipation of kinetic energy would occur at the rate comparable with

$$2\rho L\,u_*\,u_0^2 \qquad\qquad (8.17)$$

where the interfacial stress is represented by the expression $\rho u_* u_0$ in a way similar to that of Section 8.4 but with u_* somewhat larger than the $1\,\mathrm{cm\,s^{-1}}$ used there, in order to account for the ease of exchange of the air above with that in the slab. We then find that, for example, the energy is damped on a timescale of h_0/u_* or a few hours, with reasonable values for the parameters. Such simple models are interesting, and lead to a class of simple mesoscale equations, but there is another important path of progress which supposes that the evolution is continually in a slowly varying equilibrium state.

8.7 Slow disturbance about a state of no motion

For our first exploration, we take a stratified motionless atmosphere, which is disturbed by sources of heat and momentum. Equation 8.12 for continuity of mass is satisfied if we invent a stream function ψ for the x–z motion such that $u = \partial\psi/\partial z$, $w = -\partial\psi/\partial x$. Linearised about a state of no motion, equation 8.9 becomes

$$\partial v/\partial t + fu = F \qquad\qquad (8.18)$$

which shows that the momentum source serves to accelerate the air unless the Coriolis acceleration $f\,u$ intervenes. Equation 8.11 becomes

$$\partial\phi/\partial t + w\,\partial\phi/\partial z = Q \tag{8.19}$$

where $\partial\phi/\partial z = B = N^2/g$ represents the overall stratification, showing that heating makes the air warmer, except for the effect of subsidence through the term in $w\,\partial\phi/\partial z$. Mass continuity demands that the tendency to change heat and momentum are connected. Elimination of $\delta p/\rho$ from equations 8.8 and 8.10, so forming a horizontal component of the vorticity equation, then eliminating v and ϕ using equations 8.18 and 8.19 gives

$$\frac{\partial^2}{\partial t^2}\frac{\partial^2\psi}{\partial z^2} + f^2\frac{\partial^2\psi}{\partial z^2} + N^2\frac{\partial^2\psi}{\partial x^2} = f\frac{\partial F}{\partial z} - g\frac{\partial Q}{\partial x} \tag{8.20}$$

This we recognise from Section 4.3.2 as being the equation for the propagation of long internal gravity waves with forcing due to friction and diabatic heating added. For forcing such that the RHS is of the form $\exp i\,(kx\pm vz\pm\sigma t)$ the LHS contains the factor $v^2\sigma^2 - (f^2v^2 + N^2k^2)$ and we see that the time evolution will be through successive equilibrium states, at least so long as $\sigma^2 \ll f^2 + N^2\,k^2/v^2$. For typically wide shallow systems, this is true even if the forcing varies as rapidly as on a timescale of f^{-1}, so certainly for motion on the timescale of weather systems, or indeed on the diurnal cycle of sea breezes.

Another aspect of the system is illustrated by the simple steady heat source $Q = q\cos kx$. A simple particular integral of equation 8.20 is

$$\psi = -A \sin kx \quad \text{where} \quad A = gq/kN^2 \tag{8.21}$$

This is an interesting solution. We see by substituting back into equation 8.19 that the vertical advection exactly cancels out the temperature change due to the heating to give zero local change of temperature. We wonder just what we mean by the word heating in these circumstances where we put in heat energy but the system does not get warmer. The solution is not useful because it does not satisfy reasonable physical boundary conditions. Thus putting $w = 0$ at $z = 0$, demands an additional term in ψ to give

$$\psi = -A\,(1 - \exp -v_0 z)\,\sin kx \quad \text{where} \quad v_0 = Nk/f \tag{8.22}$$

which does. Because v_0 represents the decrease with height of the forced motion we see that the heating now changes the temperature in a layer of depth about f/kN near the ground where vertical displacement is inhibited. Transverse motion is also confined to this layer and serves to keep the thermal wind satisfied. Similarly, we expect that the effect of momentum sources would be nullified by horizontal motion, except where that produced convergence and consequent vertical motion which would be inhibited by the ground.

8.8 Continually balanced motion

The approximation of slow evolution suggests that we might leave out the term in Du/Dt in equation 8.8. The system then demands continual geostrophic wind balance for the x-component of the pressure gradient, or more conveniently, thermal wind balance. We can now afford to keep in other terms, and choose the non-linear terms of advection of the 'large' y-component of velocity and of ϕ. We then get

$$\frac{\partial v}{\partial t} + u\frac{\partial v}{\partial x} + w\frac{\partial v}{\partial z} + f u = F \tag{8.23}$$

$$\frac{\partial \phi}{\partial t} + u\frac{\partial \phi}{\partial x} + w\frac{\partial \phi}{\partial z} = Q \tag{8.24}$$

$$f\frac{\partial v}{\partial z} = g\frac{\partial \phi}{\partial x} \tag{8.25}$$

Now what shall we do with this set? We can use equation 8.25 to eliminate $\partial \phi/\partial t$ and $\partial v/\partial t$ from equations 8.23 and 8.24, leaving us a more complete version of equation 8.20:

$$(f^2 + f\frac{\partial v}{\partial x})\frac{\partial^2 \psi}{\partial z^2} - 2g\frac{\partial \phi}{\partial x}\frac{\partial^2 \psi}{\partial x \partial z} + N^2\frac{\partial^2 \psi}{\partial x^2} = f\frac{\partial F}{\partial z} - g\frac{\partial Q}{\partial x} \tag{8.26}$$

Apart from the modification to the first term, there is now an additional term in $\partial^2 \psi/\partial x\partial z$. This serves to slope the motion away from the vertical. For the simple heat source $Q = q \cos kx$ it is convenient to work with the complex variable $Q = q \exp ikx$ and to take the real part as solution later. Again a simple particular integral of equation 8.26 is

$$\psi = i A \exp ikx, \quad \text{where } A = gq/kN^2 \tag{8.27}$$

which, as with equation 8.21 represents the vertical velocity that just cancels out all the temperature change due to the heating. The solution satisfying the more realistic boundary conditions, $w = 0$ at $z = 0$, is

$$\psi = i A (1 - \exp i\nu z) \exp ikx$$

where

$$\nu = k(M^2 \pm i(N^2 ff' - M^4)^{1/2})/ff'$$
$$M^2 = g\, \partial\phi/\partial x \qquad f' = f + \partial v/\partial x \tag{8.28}$$

We notice that the new term in $\partial \phi/\partial x$ contribute a non-zero real part to ν and that this tends to tilt the motion away from the vertical in the sense of towards the sloping isentropic surfaces. However it is *not* the isentropic surface because the term in $\partial \phi/\partial z$ does not occur. Substituting from equation 8.25, we find that the slope is such that $f'\, dx + (\partial v/\partial z)\, dz = 0$, which is the same as demanding that $xf + v$ be constant. We recognise this number as representing something like angular momentum relative to non-rotating coordinates. We notice that this quantity also occurs in the simple 2D analysis of the sea breeze equation 8.14. For typical synoptic values the imaginary part of v is nearly v_0, and $f' \simeq f$. Indeed it turns out that the system is unstable unless

$$N^2 ff' > M^4 \text{ and } f'/f > 0 \qquad (8.29)$$

Again the heating changes the temperature only in a layer of depth f/kN near the boundary.

8.9 Evolution of the mean flow

Now we notice that as time proceeds v and ϕ will develop terms in exp ikx so the solutions will acquire terms of successively higher spatial frequency, whose vertical penetration will be even less. Perhaps a suface catastrophe impends? In fact we are interested in the modification to the flow quite as much as in the adjustment circulation itself. We can gain some crude insight into what might happen. Substituting equation 8.28 into equation 8.24 gives an equation for the rate of change in ϕ

$$
\begin{aligned}
\frac{\partial \phi}{\partial t} &= q\left(1 - \frac{v\, \partial \phi/\partial x}{k\, \partial \phi/\partial z}\right) \exp i(kx + vz) \\
&\simeq q\left(1 - i\frac{N}{f}\frac{\partial \phi/\partial x}{\partial \phi/\partial z} - \frac{g}{f^2}\frac{(\partial \phi/\partial x)^2}{\partial \phi/\partial z}\right) \exp i(kx + vz) \qquad (8.30)
\end{aligned}
$$

On the synoptic scale, the second term is of order 10^{-1}, the third 10^{-2}, but both are more nearly of order unity in frontal zones. Now we might interpret equation 8.30, as telling us something about the change in $\partial \phi/\partial x$, especially rather near the ground. Differentiating with respect to x gives, after dismantling the complex number notation,

$$\frac{\partial}{\partial t}\left(\frac{\partial \phi}{\partial x}\right) = -qk\left(1 - \frac{g}{k^2}\frac{(\partial \phi/\partial x)^2}{\partial \phi/\partial z}\right) \sin kx + qk\frac{N\, \partial \phi/\partial x}{f\, \partial \phi/\partial z} \cos kx \qquad (8.31)$$

With some stretch of the imagination we can visualise this equation as applied locally to give the local variation in $\partial \phi/\partial x$. If we do this, we notice regions near $-\pi/2 < kx < +\pi/2$ where the temperature gradient increases exponentially with time, perhaps not too surprising. What is more surprising is the possibility of a region where $\sin kx$ was negative and the gradient might behave like

$$\partial/\partial t\, (\partial \phi/\partial x) = \alpha((\partial \phi/\partial x)^2$$

or

$$\partial\phi/\partial x = 1/(c - \alpha t) \text{ where } \alpha = qkg/f^2(\partial\phi/\partial z) \tag{8.32}$$

for that suggests a catastrophic development.

Putting $\partial\phi/\partial x = 3\,\text{K}/100\,\text{km}$, $q = 10\,\text{K}/\text{day}$ gives a timescale of some 3 days for such a discontinuous temperature gradient to arise. To attack this problem directly, we might eliminate u and w between equations 8.23, 8.24 and mass conservation. During this elimination we notice Jacobian terms like $\partial(fx + v,\ \phi)/\partial(x, z)$ which suggests that a transformation of coordinates from (x, z) to $(fx + v,\ \phi)$ might be useful. Such a device puts a lot of strain on the boundary conditions which we have seen are very important, but leads to great mathematical elegance and the 'semi-geostrophic' equation set.

8.10 More general forcing

It is interesting to see how deformation changes the fields of temperature and wind, in particular with regard to thermal wind balance. Thermal wind relates the horizontal gradient of temperature to the vertical shear of the wind. We therefore examine the horizontal variation of the thermodynamic equation and the vertical variation of the momentum equation. Two-dimensionality is ensured if we suppose that the y-axis lies along the isotherms so that, $\partial\phi/\partial x$ balances $\partial v/\partial z$ and $\partial u/\partial z = 0$. Then

$$\frac{\partial}{\partial x}\left(\frac{D\phi}{Dt} - Q\right) = 0 \tag{8.33}$$

gives the rate of change of $\partial\phi/\partial x$ following a parcel. And

$$\frac{\partial}{\partial z}\left(\frac{Dv}{Dt} + fu + \frac{\partial}{\partial y}\left(\frac{\delta p}{\rho}\right)\right) - F = 0 \tag{8.34}$$

gives the rate of change of $\partial v/\partial z$ following a parcel. Carrying through the differentiations we find

$$\frac{D}{Dt}\left(\frac{\partial\phi}{\partial x}\right) = -\frac{\partial u}{\partial x}\frac{\partial\phi}{\partial x} + \frac{\partial Q}{\partial x}$$

and

$$\frac{D}{Dt}\left(\frac{\partial v}{\partial z}\right) = -\frac{\partial v}{\partial z}\frac{\partial v}{\partial y} + \frac{\partial F}{\partial z} \tag{8.35}$$

and, combining the two

$$\frac{D}{Dt}\left(\frac{\partial v}{\partial z} - \frac{g}{f}\frac{\partial\phi}{\partial x}\right) = \frac{\partial F}{\partial z} - \frac{g}{f}\frac{\partial Q}{\partial x} - \frac{\partial v}{\partial z}\left(\frac{\partial u}{\partial x} - \frac{\partial v}{\partial y}\right) \tag{8.36}$$

Equation 8.36 shows the particular combination of the effects of stress, heating and deformation that disturb thermal wind equilibrium, and therefore generalises the RHS forcing term in equation 8.26. We notice that though the three influences are probably of similar size, according to equation 8.35 the effect of deformation is likely to increase exponentially in time. Thus, for the deformation,

$$\partial v/\partial z \propto \exp\left(\partial v/\partial y - \partial u/\partial x\right) t \tag{8.37}$$

so the deformation process is likely to be exponential. Heating has a similar exponential quality when the heating increases with vertical velocity, as one might expect with wet convection. Friction at the surface is likely to have some damping influence, but where it serves to generate vertical diplacements releasing latent heat, the overall effect may yet be destabilising. With all these destabilising influences it is not unexpected that there should be such a great range of mesoscale phenomena such as those displayed on satellite pictures. Finally, equation 8.37 shows the convergence of isotherms visualised by Bergeron in another light. The term in $\partial\phi/\partial x$ in the thermal part shows that if there is convergence of the isotherm field, $\partial u/\partial x$ is negative, and there will indeed be an exponential increase in the temperature gradient. However, the horizontal wind is nearly non-divergent, so $\partial v/\partial y$ must be positive, so the momentum equation demands that the shear *decreases* exponentially with time. That this must be so is easily reconstructed, but demands the acceptance that the deformation field carries the shear as well as the shear carrying the deformation.

Chapter 9

Motion of large scale

9.1 Introduction

When the velocities of particles change slowly with time the geostrophic approximation to the horizontal wind becomes more accurate. While at first sight this seems a good thing, because it makes the equations simpler, but it is also worrying, for the momentum equations lose some of their ability to be predictive. Thus the term in Dv/Dt is a small residual between two large terms. If horizontal gradients of pressure are largely balanced by the Coriolis force, how are we to find what is left over to make the momentum evolve? This dilemma can be tackled by eliminating the pressure term from the momentum equations. This gives, of course, the vertical component of the vorticity equation.

9.2 Scale analysis

We use the equations of motion to establish a set of order-of-magnitude relations between variables. This process is analogous to solving the equations, except that it has the lesser aim of seeing that the numbers involved could possibly represent a solution. While we aim to be deductive, we find that it is easier to justify some approximations only after they have been made. Thus our end point is really one of plausible consistency.

Many weather systems have a much longer transverse than longitudinal scale. This is consistent with the notion that they are there in order to transfer properties like heat in the transverse direction. The evidence, as illustrated in

Figure 1.5, is most marked in the trajectories and in the flow relative to the system, and least clear in the flow relative to the surface which is the picture we are usually given by the synoptic chart and with which we are most familiar. Here we take advantage of this transverse nature of the scales by taking the dominant wavelength to be in one dimension, which we will call x, and identify with the W-E direction on the Earth. Thus k is the wavenumber in the x-direction, defining wavelength $2\pi/k$. Similarly there is a vertical scale H. Thus $\partial/\partial x \sim k$, $\partial/\partial z \sim 1/H$.

Let

- V be a typical horizontal velocity, measured relative to the wave
- W a typical vertical velocity
- $\Delta\phi$ the magnitude of the horizontal variation of ϕ
- Δp the magnitude of the horizontal variation in pressure
- $B = \partial\phi/\partial z$ the static stability.

These quantities are not all independent of each other and, by analogy with solving the equations of motion to find their detailed relation, we here use the equations to describe the magnitude of this relation. We assume that the patterns are advective in character. What we mean by this is that the patterns develop rather slowly compared with the time taken for air to pass through them. This 'advective' assumption gives (in contrast to the shearing instability of Section 6.1) $D/Dt_h \sim kV$. It follows that the ratio of relative acceleration to Coriolis acceleration is kV^2/fV, where we have assumed that the vertical advection of momentum is not greater than the horizontal. This ratio generates a Rossby number

$$Ro = kV/f \tag{9.1}$$

whose smallness ensures geostrophic balance. Assuming that geostrophy is at least a crude approximation, we can argue that the thermal wind and the shear are of the same size, $\partial v/\partial z \sim (g/f)\mathbf{k} \wedge \nabla\phi$ so

$$\Delta\phi \sim fV/gHk \tag{9.2}$$

which connects two of the variables.

If diabatic effects are not dominant, the thermodynamic equation gives $D\phi/Dt_h \sim BW$, so

$$W \sim kV\,\Delta\phi/B \sim fV^2/N^2H \tag{9.3}$$

which connects two more. The criterion for neglect of vertical advection, coincides with that for the neglect of $\partial w/\partial z$ compared with either $\partial u/\partial x$ or $\partial v/\partial y$ (but not their sum) in the equation of continuity of mass, and is

$$\frac{w\,\partial/\partial z}{\mathbf{v}.\nabla_h} \sim \frac{\partial w/\partial z}{\partial u/\partial x} \sim \frac{\Delta\phi}{BH} \ll 1 \tag{9.4}$$

This can be interpreted as saying that hardly any air goes from top to bottom of the system, which is well satisfied for most of the air on the synoptic scale, but we notice that the small mass of air which does go from bottom to top mainly in the frontal zones has a profound effect on the weather.

Similarly, the term in $w\,\partial\rho/\partial z$ in the equation of continuity of mass can be neglected if

$$\Delta\phi/BH_0 \ll 1 \qquad (9.5)$$

where H_0 is the density scale height. The variety of possible systems can be further restricted through reference to the vorticity equation. The vertical component of the vorticity equation, as in Section 3.12, has terms of advection, tipping, and stretching, of absolute vorticity.

We hope that the advection of relative vorticity $D\zeta/Dt_h \sim k^2 V^2$ is significant, since this term is the sole representative of the predictive ability of the equations, which is what we are trying to isolate. We also expect vertical advection to be negligible, in view of inequality 9.4, but advection of Earth's vorticity $Df/Dt \sim \beta V$ not necessarily so.

The vector vorticity has horizontal components of order V/H, and vertical component kV, so in the 'stretching-tipping' term

$$(f\mathbf{k} + \operatorname{curl}\mathbf{v}).\nabla \sim (f + kV)/H + (f + V/H)k \qquad (9.6)$$

with $f \gg kV$ from equation 9.1, and $f \ll V/H$ well satisfied in the troposphere, the vertical stretching of planetary vorticity is more important than the tipping terms. Thus for these three remaining terms in the vorticity equation

$$\frac{D\zeta_z}{Dt} \sim k^2 V^2 \qquad \frac{Df}{Dt} \sim \beta V \qquad f\frac{\partial w}{\partial z} \sim \frac{f^2 V^2}{N^2 H^2} \qquad (9.7)$$

9.3 Simplified equations

Omitting the terms shown to be negligible above, the equations of motion can be simplified. The vertical component of the vorticity equation, neglecting the tipping and stretching of relative vorticity, and vertical advection of absolute vorticity, is

$$\frac{D}{Dt_h}(\zeta_z + \beta y) = \frac{f}{\rho_0}\frac{\partial}{\partial z}(\rho_0 w) \qquad (9.8)$$

We compare the terms in equation 9.8 using approximations 9.7. If the first and last are comparable, then $k^2 H^2 \sim f^2/N^2$. This is Eady's relation (1949) for the wave that amplifies most rapidly in a baroclinic fluid of depth H. Moreover, identifying the characteristic value V with the variation in the unperturbed flow ΔU gives

$$Ro = kV/f = \Delta U/HN = Ri^{-1/2} \tag{9.9}$$

This rather neat result, shows that if the Richardson number of the mean flow is large, then the eddy motion which develops spontaneously on it will have a small Rossby number and therefore be quasi-geostrophic. Criterion 9.4 becomes $\Delta\phi/BH = Ri^{-1/2} \ll 1$, and condition 9.5 becomes

$$H \, Ri^{-1/2} \ll H_0 \tag{9.10}$$

The ratio of the exponential to advected timescales is about 0.4 according to Eady's development, so the advective assumption is retrospectively adequate.

If the first and second terms in the vorticity equation are comparable, we get the Rossby-wave dispersion relation. Equating the second and third, which is the non-predictive set, gives $H_*^2 = f^2 \Delta U/N^2 \beta$, which is Charney's relation for the upward penetration distance (H_*) of a quasi-geostrophic wave.

The thermodynamic equation with the diabatic source term Q included is

$$\frac{D\phi}{Dt} = Q \simeq \frac{D}{Dt_h} \delta\phi + Bw \tag{9.11}$$

and continuity of mass

$$\partial u/\partial x \simeq - \partial v/\partial y \tag{9.12}$$

Notice how much better this statement is than $\partial u/\partial x + \partial v/\partial y \simeq 0$, because we know the almostness of the approximation is qualified by the size of either term in expression 9.12. A stream function ψ for the horizontal component of the motion can be introduced, where

$$u \simeq -\partial\psi/\partial y \qquad v \simeq +\partial\psi/\partial x \tag{9.13}$$

and, somewhat incidentally,

$$\delta p \simeq \rho_0 f \psi \tag{9.14}$$

We notice that the stream function is for flow in the horizontal plane, rather than the vertical plane as in Chapters 5 and 8.

From the horizontal component of the vorticity equation we have the thermal wind which gives a similar set

$$\frac{\partial u}{\partial z} = -\frac{g}{f}\frac{\partial\delta\phi}{\partial y} \qquad \frac{\partial v}{\partial z} = +\frac{g}{f}\frac{\partial\delta\phi}{\partial x}$$

where

$$\delta\phi = \frac{f}{g}\frac{\partial\psi}{\partial z} \tag{9.15}$$

Eliminating $\delta\phi$ between relations 9.15 and 9.11, gives an expression for w in terms of the non-divergent stream function.

$$w = \frac{Q}{B} - \frac{f}{gB} \frac{D}{Dt_h} \left(\frac{\partial \psi}{\partial z} \right) \tag{9.16}$$

This equation is remarkable in the sense that it gives the vertical velocity (using the thermodynamic equation) in terms of a function whose derivation demanded that there was negligible vertical velocity, but this time in the equation of continuity of mass. A nice illustration of the neglect of a term being dependent on what the term is to be used for. Again we notice the approximation $w \simeq Q/B$ which generates motion that just cancels out the local temperature change due to the heating, as in Section 8.7. Substituting the value for w given by equation 9.16 into equation 9.8 gives the quasi-geostrophic form of the vorticity equation

$$\frac{D}{Dt_h}(\zeta_z + \beta y) = \frac{f}{\rho_0} \frac{\partial}{\partial z} \left(\frac{\rho_0}{B} Q \right) - \frac{1}{\rho_0} \frac{\partial}{\partial z} \frac{D}{Dt_h} \left(\frac{\rho_0 f^2}{gB} \frac{\partial \psi}{\partial z} \right)$$

$$= \frac{f}{\rho_0} \frac{\partial}{\partial z} \left(\frac{\rho_0}{B} Q \right) - \frac{1}{\rho_0} \frac{D}{Dt} \left(\frac{\partial}{\partial z} \frac{\rho_0 f^2}{gB} \frac{\partial \psi}{\partial z} \right)$$

whence

$$\frac{D}{Dt_h} \left(\zeta_z + \beta y + \frac{1}{\rho_0} \frac{\partial}{\partial z} \left(\rho_0 \frac{f^2}{N^2} \frac{\partial \psi}{\partial z} \right) \right) = \frac{1}{\rho_0} \frac{\partial}{\partial z} \left(\frac{f}{B} Q \right)$$

where

$$\zeta_z = \partial^2 \psi / \partial x^2 + \partial^2 \psi / \partial y^2 \tag{9.17}$$

This set of equations forms a useful starting point for the study of large-scale motion. It is a forgiving set. For example, if the atmosphere were barotropic, in the sense of having no horizontal temperature gradients, then the derivation can still be carried through, though some steps may now be redundant.

9.4 Potential vorticity

From equation 9.17, we see that in adiabatic frictionless flow, the quantity

$$\left(\zeta_z + \beta y + \frac{1}{\rho_0} \frac{\partial}{\partial z} \left(\rho_0 \frac{f^2}{N^2} \frac{\partial \psi}{\partial z} \right) \right) \tag{9.18}$$

is conserved by fluid particles whose vertical displacement is, and can be, ignored. This quantity is closely related to potential vorticity. However, true potential vorticity, sometimes called Ertel-potential vorticity, has a large vertical variation, and its vertical advection cannot be ignored. The quasi-geostrophic vorticity equation 9.17 can be derived from that for conservation of true potential vorticity by treating the horizontally averaged potential vorticity and its vertical advection separately. It is largely a matter of taste whether to use the more accurate form of potential vorticity, which is also less tractable, than the simpler but less accurate quasi-geostrophic form.

9.5 The parcel theory of baroclinic instability

We need some test cases in order to understand what these equations imply. For example, they should be able to describe stable Rossby waves propagating through a barotropic atmosphere, and the instability of a statically stable atmosphere with a horizontal gradient of temperature, when the motion takes place in surfaces inclined to the horizontal.

Suppose the isentropic surfaces slope as shown in Figure 9.1, with potentially warmer air above, and for smaller values of y. Now imagine that two parcels of air are exchanged adiabatically. If the path of exchange is along the vertical, then the final potential energy will be greater than the original, and the exchange will not occur spontaneously. If the exchange is along the isentropic surface the potential energy does not change, because the two parcels have the same potential temperature. If the exchange is along the horizontal, the potential energy is unchanged because the height of no particle is changed. Perhaps, in the narrow range between these two, potential energy may be released to drive convection which is essentially slantwise. Thus argued Eady.

The argument may be rationalised to show that, for a fixed displacement, the maximum potential energy is released when the parcel paths have a slope

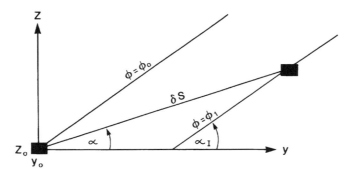

Fig. 9.1 Distribution of potential temperature on a vertical plane across the isentropes showing the consequences of parcel exchange along some sloping paths. This is the essence of baroclinic instability visualised as slantwise convection. Though the air that goes up is potentially warmer than the air that goes down remember that the motion is adiabatic, so the isentropic surfaces are carried along with the fluid. Thus, the isentropic surface, initially flat, becomes distorted into corrugations with the cold air penetrating equatorward at the ground, warm air polewards at the top.

half-way between the original isentropes and the horizontal. Calculation of available potential energy gives

$$2\left(\frac{1}{2} m\mathbf{v}^2\right) \simeq -m g\,\Delta z\,\Delta\phi \simeq m g\,(\Delta y)^2 (\partial\phi/\partial y)^2/4B \qquad (9.19)$$

In this calculation we assume that no other parcel acquires any of the potential energy released, as they might by being pushed aside by the pressure field for example. Equation 9.19 gives the parcel speed $|\mathbf{v}|$ from

$$|\mathbf{v}|^2 \simeq g\,(\Delta y)^2 (\partial\phi/\partial y)^2/4B \qquad (9.20)$$

The amplification rate is a useful way of evaluating the efficiency of an instability process. Suppose that all the kinetic energy is taken by the y-component of velocity. Then the LHS is just $(D\,\delta y/Dt)^2$, and the amplification rate can be found from

$$\frac{1}{\delta y}\frac{D}{Dt}\delta y = \frac{1}{2}\left(\frac{g}{B}\right)^{1/2}\left(\frac{\partial\phi}{\partial y}\right) \qquad (9.21)$$

Substituting suitable numerical values ($B \sim 10^{-5}\,\text{m}^{-1}$, $\partial\phi/\partial y \sim 10^{-7}\,\text{m}^{-1}$) gives a doubling time of some 1 or 2 days.

9.6 Simplest baroclinic wave

Guided by these considerations we can begin to put together a dynamical theory of such slantwise convection. Eady, in his classic study, supposed that no essential features would be lost if the atmosphere were supposed to be incompressible in dry adiabatic motion with no variation of Coriolis parameter with latitude, and no variation of static stability (B) up to a rigid tropopause. Thus he put $Q = \beta = d\rho_0/dz = 0$. If in addition, the initial flow has constant shear $U = U_0 + A z$, then $\nabla^2\psi + (f^2/N^2)\,\partial^2\psi/\partial z^2$ vanishes everywhere initially, and must, according to equation 9.17, remain zero for each and every parcel thereafter. Solutions that fit into a channel of width π/μ, can be written

$$\psi = Bz - y(U_0 + Az) + F(z)\,\exp\,\mathrm{i}k(x - ct)\,\cos\,\mu y$$

where $F(z)$ satisfies

$$\partial^2 F/\partial z^2 - v^2 F = 0, \quad \text{and } v^2 = (N/f)^2(k^2 + \mu^2)$$

so

$$F = a\,\cosh\,vz + b\,\sinh\,vz \qquad (9.22)$$

From equation 9.16, the condition $w = 0$ at the horizontal lids demands

$$D/Dt_h(\partial\psi/\partial z) = 0 \qquad (9.23)$$

This far, the analysis is completely non-linear but we concentrate here on the linearised evolution, so suppose that F is sufficiently small that all quadratic terms in equation 9.23 can be neglected. This makes the boundary condition

$$\left(\frac{\partial}{\partial t} + U\frac{\partial}{\partial x}\right)\frac{\partial \psi'}{\partial z} + v\frac{\partial}{\partial y}\left(\frac{\partial \psi_0}{\partial z}\right) = 0$$

or

$$(U_0 + zA - c)\frac{dF}{dz} - AF = 0 \tag{9.24}$$

at the lids. The algebra is eased a little if we choose $z = 0$ as the middle level, which makes U_0 the mean speed in the layer. We find that F can satisfy the boundary conditions only if

$$(c - U_0)^2 = H^2 A^2 \left(\frac{1}{X^2} + \frac{1}{4} - \frac{\coth X}{X^2}\right) \text{ where } X = vH \tag{9.25}$$

For each value of X there are two solutions to equation 9.25. For waves so short that $X \gg 1$, $c \sim U_0 \pm HA/2$, showing that these short waves are neutral in the sense that they do not amplify with time, and that they move with the velocity near that of the fluid at either the top or the bottom boundary.

For waves so long that $X \ll 1$, we find from equation 9.25 $c \simeq U_0 \pm iHA/\sqrt{12}$. These form an amplifying/diminishing pair, each moving with the velocity of the mean wind. Further calculation shows that the maximum amplification rate for variation in μ is for $\mu = 0$; emphasising the large scale of the transverse motion, and has $X = 1.61$ nearly, which gives

$$kc_i = 0.310 \, Af/N \tag{9.26}$$

We notice that the ratio of advective to growth timescales $kc_i/k(U - c)$ is $0.31H/1.61z$, for $-H/2 < z < H/2$, so of size unity, but emphasising the difference in particles that are, and are not, trapped at the steering level of the motion.

By solving for a and b in equation 9.22, we can find the shape of the amplifying wave, by which we mean the phase, as well as the amplitude relations; like whether the warm air is going up and the cold down. The real and imaginary parts of the solution hold information about both the amplitude and the x-phase of the different functions. Figure 9.2 shows the relative phases of temperature, upward, and poleward components of velocity, while Figure 9.3 shows the motion on an isentropic sheet. The isentropic surfaces at $z = 0$ and $z = H$ are being advected along these surfaces to become more corrugated. This picture is rather close to Ludlam's reconstruction of the flow in a mature mid-latitude depression shown in Figure 1.4(c). The flow is not easy to visualise because the motion is amplifying with time, as well as advecting. Real systems are likely to have a short sharp amplifying phase during which the characteristic

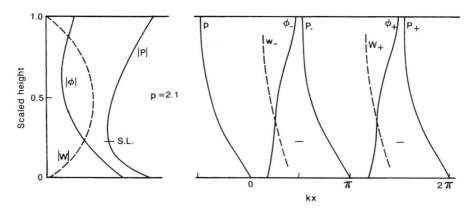

Figure 9.2 Cross-section along the plane perpendicular to that of Figure 9.1 constructed from a calculated solution to the baroclinic instability problem. Notice the good correlation between vertical velocity and poleward velocity and temperature, showing efficient polewards, and upwards transport of energy. For the typical values of β assumed here, the steering level is suitably depressed as expected from the additional Rossby-wave retardation.

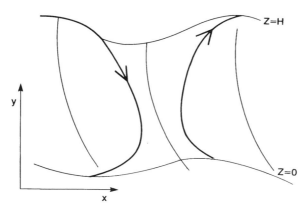

Figure 9.3 Streamlines relative to the wave of the motion shown in Figure 9.2. The isentropic surfaces at $z = 0$ and $z = H$ are being advected along these surfaces to become more corrugated. The solution is not easy to visualise because the motion is amplifying with time, so the trajectories and streamlines do not coincide. The analysis of Figure 1.4 shows some similarity. Real systems have a rather short amplifying phase in which the characteristic wavelength and shape are selected, followed by a mature phase in which the majority of the heat transfer takes place, finally dying rather slowly through dispersion or dissipation. Section 14.2 is not inconsistent with parts of this picture.

wavelength and shape is selected, followed by a mature phase in which the majority of the heat transfer takes place, finally dying out rather slowly by dispersion and dissipation during which phase the momentum transfer takes place. Section 14.2 is not inconsistent with this picture.

Comparing the amplification rate given by equation 9.26 with the ideal thermodynamic solution of equation 9.20 we see that the constraints imposed by the equations of motion and by the boundary conditions have reduced the amplification rate from 0.5 to 0.310. We can even persuade our mathematically correct dynamical solution to become more like the efficient solution by the device of tilting the boundaries towards the optimum path.

9.7 Sloping boundaries

If both the rigid boundaries slope at angle ϵ to the horizontal, then the linearised boundary conditions become $w = \epsilon v$ at $z = \pm H/2$. Eliminating a and b much as before, we find the frequency equation

$$v^2 T (\Delta^2 - c^2) - v \Delta U_* (T^2 + 1) + U_*^2 = 0 \qquad (9.27)$$

where

$$
\begin{aligned}
T &= \tanh vH/2 \\
U_* &= A - gB\epsilon/f \\
\Delta &= HA/2
\end{aligned}
\qquad (9.28)
$$

then a little more manipulation gives

$$\left(\frac{c}{AH}\right)^2 = \frac{1}{4} + \frac{(1-\gamma)^2}{v^2 H^2} - \frac{(1-\gamma)}{vH}\coth vH \qquad (9.29)$$

where $\gamma = (H/A)(N^2/f)\epsilon$, is the slope of the boundary divided by the initial slope of the isentropes.

With $\gamma = 0$ we recover Eady's solution, equation 9.25. With $\gamma = 1$ we find that the perturbed flow is along the isentropic surfaces everywhere and cannot make use of the potential energy for any wavelength. It is not obvious to me why this should be. As γ increases from zero, the wavelength of maximum growth rate also increases, with finite growth rate $kc_i = A(\gamma(1-\gamma))^{1/2}$ for $\mu = 0$ and $k \to 0$ and maximum for $\gamma = 1/2$ which gives the boundaries half the isentropic slope for flow of unbounded wavelength, for which values the full exact solution 9.29 recovers the optimum growth rate of the parcel solution, so long as the x-wavelength is very large.

9.8 The slantwise nature of the convection

The parcels of air engaged in this instability largely follow paths confined to a plane sloping at half the slope of the initial isentropic surfaces. Let us exploit this feature. The vector vorticity equation consistent with the approximations of this chapter is essentially equation 3.25 with the hydrostatic approximation for ∇p

$$\mathbf{k}\frac{DZ}{Dt}+f\frac{\partial \mathbf{v}}{\partial z}-g\mathbf{k}\wedge\nabla\phi=0 \qquad (9.30)$$

where Z is the vertical component of absolute vorticity, and $\mathbf{v}$ is the three-dimensional velocity. We might make a model of baroclinic motion by supposing that it took place completely in a shallow sheet inclined to the horizontal at shallow slope α, intended to represent the half-isentropic plane. Because the sheet is shallow there is negligible component of velocity in the perpendicular direction. Thus the motion in the plane must be non-divergent, and we can use a stream function to describe it. There is no vortex stretching term in equation 9.30 but there is a component of the baroclinic term, which is $g \sin \alpha\,(\partial \phi/\partial x)$. The thermodynamic equation again demands conservation of ϕ but now in the tilted plane. We notice that in this plane, ϕ is initially larger in the equatorward (lower) part of the system, compared with the poleward (upper) part. In fact the equations now look exactly the same as those for the two-dimensional convective instability of Section 7.6, driven by the temperature contrast along the sloping suface, and with the acceleration due to gravity replaced by $g \sin \alpha$. Boundary conditions of no flow through the boundaries are the same as for the vertically overturning system too.

Perhaps this is why Ludlam's simple cumulonimbus circulation of Figure 1.6(d) which does not look like a cumulonimbus, looks very much like the trajectories in a quasi-geostrophic wave. We notice that the orientation of the motion is such as to transfer momentum polewards. So here is yet another hypothesis for the poleward momentum transfer observed in baroclinic waves.

9.9 More general stability problems

The constraints of setting β and H_0 to zero and imposing rigid lids at top and bottom can be relaxed, but only at the expense of greatly increasing the mathematical labour needed to find solutions. What were rather simple exponential functions become confluent hypergeometric functions. Bessel functions of complex argument occur even in simple limit cases, while Bessel functions of complex order are not unheard of. The labour involved in manipulating such trancendental functions, which often has to be done numerically, is so great that it is easier to solve the differential equations numerically from the start. However, simple results can be obtained in some limits, like comparatively long and comparatively short waves.

An important effect arises when the variation of Coriolis parameter with latitude β is taken into account. As might be anticipated from the behaviour of barotropic waves, this is most important for longer waves and tends to retard the phase speed like β/k^2, i.e. the Rossby phase speed. It is found that if the retardation is so large as to take the steering level outside the fluid then the wave becomes neutral. For very short waves β has a curious effect. In Eady's problem, with $\beta = 0$, short waves form a neutral pair, with steering level near either the upper or the lower boundary. When we make β appropriately small and positive, the upper wave ceases to exist, and the lower wave becomes a feebly amplifying/diminishing pair. This apparent discontinuity in physical properties is rather worrying, until one realises that it is the mathematical device of looking for solutions of constant shape (i.e. separable in time and space) that is responsible. Thus the upper wave which exists for $\beta = 0$ begins to change its shape very slowly when β is positive and so disappears from the category of fixed-shape waves. The physical wave still exists and is little different from what it was before, except that it fails to keep its shape exactly independent of time. This notion is germinal and is explored further in Section 10.2.

9.10 Short waves

Very short waves sample small regions of space, and detect places favourable for energy conversion. The one-dimensional stability problem illustrates this nicely. We choose to include variation of U with z, and non-zero values of β, both of which have interesting effects. We ignore the variation of inertial density with height because it creates mainly a distortion largely described by the $\rho^{1/2}$ effect and slightly modifies the effect of β. With this approximation the linearised equation is

$$(U - c)\left(M^2 \frac{\partial^2}{\partial z^2} - k^2\right)\psi + \left(\beta - M^2 \frac{\partial^2 U}{\partial z^2}\right)\psi = 0 \tag{9.31}$$

where $M = f/N \sim 10^{-2}$ is a factor representing the ratio of vertical to horizontal scales. The boundary condition at rigid lids, as in equation 9.23 demands

$$(U - c)\frac{\partial \psi}{\partial z} - \frac{dU}{dz}\psi = 0 \tag{9.32}$$

When k is suitably large the solution to equation 9.31 must be exponential almost everywhere. The exception being for values of z near the steering level, where $(U - c)$ becomes small. But if the imaginary part of c is small we can integrate over this singularity as if it were in the complex plane to join the two solutions from opposite sides. Expanding about the steering level $z = z_*$, defined by $U(z_*) = c_0$, the real part of the phase speed, we have

$$M^2 \frac{\partial^2 \psi}{\partial z^2} \simeq -\frac{(\beta - M^2 U'')}{(z - z_*)U'_* - ic_1}\psi$$

so that

$$\left(\frac{\partial \psi}{\partial z}\right)^{z_*+} - \left(\frac{\partial \psi}{\partial z}\right)^{z_*-} \simeq \pm i\pi \frac{(\beta - M^2 U_*'')}{M^2 U_*'} \psi \tag{9.33}$$

as the conditions to be satisfied near to, and as the steering level is bridged, respectively. The sign ambiguity is determined by the position of the singularity. Suppose that we seek the amplifying wave, which therefore has the imaginary part of the phase speed c_i positive. Then if U_*' is positive, the singularity is above the real axis and integration along the real axis navigates the singularity in the mathematically conventional positive sense to give the upper sign. If U_*' is negative the singularity is on the other side of the real axis, and the lower sign is appropriate. Thus, whatever the sign of U_*', the contribution to the integral is in the sense $+i\pi/|U_*'|$.

We can fill in the rest of the solution rather simply. The steering level must be a non-zero distance away from one boundary. Outside this region the solution, which is like $\exp kz/M$, must be proportional to the exponential that decreases away from the boundary if it is to remain bounded as $k \to \infty$. It is convenient to introduce the stretched wavenumber; $K = k/M$. For a wave with steering level near the lower boundary $z = 0$

$$\psi = \alpha \exp -Kz \quad \text{for } z > z_*$$

and

$$\psi = a \cosh Kz + b \sinh Kz \quad \text{for } 0 < z < z_* \tag{9.34}$$

and we must have

$$K(U_0 - c)b = U_0' a \tag{9.35}$$

to satisfy the condition at the lower lid. At the steering level $z = z_*$

$$(\psi'/\psi)_+ = -K$$

and

$$(\psi'/\psi)_- = K(a \sinh Kz_* + b \cosh Kz_*)/(a \cosh Kz_* + b \sinh Kz_*) \tag{9.36}$$

Thus, $[\psi'/\psi]_-^+ = -K/(1 + (a-b)/(a+b)) \exp -2Kz_*$ and equating this with the change obtained for integration round the singularity of equation 9.33 gives

$$1 + \frac{a-b}{a+b} \exp -2Kz_* = -\frac{iK}{\pi} \left(\frac{M^2 |U_*'|}{\beta - M^2 U_*''}\right) \tag{9.37}$$

Finally, equations 9.35 and 9.37 together determine $c(k)$, and $a(k)/b(k)$. Since z_* is real by definition, the RHS of equation 9.37 must be purely imaginary, and can be balanced only by an imaginary part of $(a-b)/(a+b)$. For K indefinitely large, the RHS of equation 9.37 varies directly with K, unless $U_*' = 0$, which is unlikely and we infer that $a + b \to 0$ like K^{-1}, which gives, from equation 9.35

$$K(U_0 - c) + U_0' + A/K = O(K^{-2}) \tag{9.38}$$

where A is to be determined, or

$$c = U_0 + U_0'/K + A/K^2 \text{ where } z_* = 1/K + O(K^{-2}) \tag{9.39}$$

and finally, from equation 9.37

$$A = 2\pi i e^{-2} \frac{(\beta - M^2 U_0'') U'}{M^2 \mid U_0' \mid} \tag{9.40}$$

The assumption that c_1 was positive demands that $U'(\beta - M^2 U'')$ evaluated near the lower boundary, be positive. Apart from this sign determination, the actual value of U' seems to be irrelevant. Near an upper rigid boundary, the same form holds except that $U'(\beta - M^2 U'')$ must be negative for instability.

This illustrates the discontinuous behaviour for the existence of short waves as the apparent value of β changes sign in Eady's solution. We might imagine a system in which the shear U' changes sign with height. Indeed this is often true in the tropics, particularly over the eastern Atlantic ocean near 20°N. Our discussion then suggests the possibility of two possible growing waves, one concentrated near the ground, and one in the upper troposphere, so long as $\beta - M^2 U''$ was positive. Figure 9.4 shows some numerical solutions for intermediate wavelengths. We can believe that the upper solution is a distorted version of the usual Eady wave, with similar value for the wavelength of maximum growth rate. The additional solution is found near the lower boundary, is shorter with wavelength proportional to the depth of the layer of opposite shear. It is also less rapidly amplifying, in proportion to the total shear of the layer.

This lower wave may be the mechanism for generating easterly propagating waves sometimes seen on satellite pictures of the tropics. The energetics of these waves is controlled by the release of latent heat not by baroclinity, indeed some of them may transform into hurricanes whose dynamics is dominated by convergence of air carrying angular momentum towards clusters of cumulonimbus. What we suggest here is that it is the large-scale organisation and spacing of the wet convection which is governed by this comparatively feeble, baroclinic processes. Maybe the processes that allow the convection to persist, which is related to the transformation to the hurricane, are also of large scale.

We might also seek solutions in which $a \sim b$ in some region, for this wave would decrease away on both sides of an internal steering level. This turns out to be abortive, for reasons which are related to the analysis of Section 6.4. For K large, the boundaries appear indefinitely far away and the condition is one of monotonous decrease away from the steering level, on both sides. As we saw in Section 6.4 very short waves cannot detect both sides of the shear zone, and this makes them stable. We might, by analogy expect longer baroclinic waves also to be unstable. This possibility is explored in the next section.

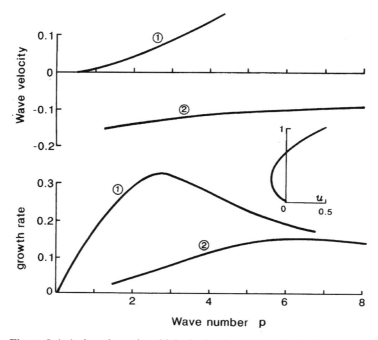

Figure 9.4 A deep layer in which the horizontal gradient of temperature changes sign with height can support two baroclinic waves at each wavelength. Here a shallow layer of easterly shear lies under a deep one of westerly shear and there are two waves. Wave 1 occupies most of the depth available and is long. Wave 2 occupies the shallow layer of easterly shear, and is shorter, proportional to the depth of the sheared layers.

9.11 Integral constraints

An interesting extension of the criteria for stability arises from an integral constraint. This is similar to one for shearing instability developed by Rayleigh. We take the equation in the form

$$M^2 \frac{\partial^2 \psi}{\partial z^2} - k^2 \psi + \frac{(\beta - M^2 U'')}{(U - c)} \psi = 0 \qquad (9.41)$$

and extract information about the imaginary part. This we can do by multiplying through by the complex conjugate ψ^*. A little further manipulation allows us to write

$$\frac{\mathrm{d}}{\mathrm{d}z}(M^2 \psi^* \psi') - M^2 \psi' \psi^{*'} + \left(\frac{\beta - M^2 U''}{U - c} - k^2 \right) \psi^* \psi = 0 \qquad (9.42)$$

All but the first and last term of this equation are purely real, and the last term has an imaginary part only if c_i does not vanish. Integrating the equation through the depth of the fluid, the first term sometimes vanishes. This happens if ψ decreases to small values at the extremities, which happens if the wave is confined to a rather narrow height range. Another possibility is that the boundary condition, at a finite distance, might be such as to make the term vanish. This would be true if either ψ' or ψ were to vanish. The first would be true of rigid boundaries at which the unperturbed flow had vanishing shear. The second was true of the sheared flow studied by Rayleigh. In these cases c_1 can be non-zero only if

$$\int (\beta - M^2 U'') \frac{|\psi|^2}{|U - c|^2} \, \mathrm{d}z = 0 \qquad (9.43)$$

Since all the contents of the integrand, apart from the β-term are positive, it is necessary that the term in $\beta - M^2 U''$ must change sign somewhere within the range of integration for the integral to vanish. By analogy with the analysis of flow with a distributed shear in Section 6.4, it is likely that the wave may not be too short or it will not be able to appreciate the change in sign of $\beta - M^2 U''$, in much the same way that a shearing instability wave must be able to 'see' the maximum in the shear against a background of lesser shear.

Thus we visualise two types of solution. A boundary wave, perhaps of rather short wavelength, utilising the shear zone close to a boundary, and a rather longer internal wave. Eady had this sort of picture for the long waves and cyclone waves; with the long waves, of global scale wavenumber 3 to 6, concentrating the baroclinity into rather shallow frontal zones where it was available to generate 'wave cyclones' of distinctly smaller scale. Supposing that the total baroclinity available to each scale is comparable; the larger scale merely brings the contrasting air masses closer together, then, since they share the same available potential energy, we might suppose that the horizontal velocities may also be comparable. This implies that the vorticities are proportional to the wavenumber, and through the vorticity equation that the vertical velocities, and therefore the rainfall rates, would be proportional to the square of the space scale. Thus we might expect 'weather' to be concentrated into the smaller-scale systems, that rely on the longer-scale ones for their energy supply, location and persistence.

9.12 Another integral relation

The equations may be juggled in many different ways. Equation 9.41 may be written in the form

$$M^2 \frac{\mathrm{d}}{\mathrm{d}z} \left((U - c)\psi' - U'\psi \right) + \left(\beta - k^2(U - c) \right) \psi = 0 \qquad (9.44)$$

The first term vanishes on integration between rigid lids because it is proportional to the vertical component of velocity. This gives the result

$$c = \frac{\int U \psi \, dz}{\int \psi \, dz} - \frac{\beta}{k^2} \tag{9.45}$$

At first sight this looks like a useful generalisation of Rossby's formula for the wave velocity, but we find that the evaluation demands detailed specification of the form of ψ, and good guesses will not do. We also recognise that equation 9.45 is only the quasi-geostrophic vorticity equation, and the rather complicated part in the brackets merely $\partial w / \partial z$ expressed in terms of ψ. Thus, for example, there is no information about the baroclinity of the system in equation 9.45, apart from what we put into it through the function ψ.

9.13 Completeness and the complex plane

Why does the analysis keep straying into the complex plane, when we are dealing with problems in real space? The device of using $\exp ikx$ to represent suitable combinations of sines and cosines is one possibility. But we notice that all direct reference to this device has disappeared from some of our equations; as in equation 9.41, for example. Perhaps it is the complex phase velocity that has not disappeared. Thus it is the term in $\exp ikct$ that carries the implication of complex numbers. It is some indication of the lack of ordinary wave solutions to the equations.

In principle, we ought to expect to be able to solve the general linear forecasting problem; given a vanishingly small, but otherwise arbitrary, initial perturbation $\psi(x, y, z, t = 0)$, of some flow $U(z)$, the linearised equations predict the evolution of the system, at least while the perturbation remains of small amplitude.

So far, we have concentrated attention on those solutions that amplify with time, and have been diverted into the complex plane. We have found that, for a given wavelength, there are not many such solutions; often only two different functions of height at any one wavelength. This is not enough to express an arbitrary function of height so there must be many others to make up a complete set.

In the next chapter we consider what these other functions might be.

Chapter 10

The forecast problem

And ever changing,
like a joyless eye
That finds no object worth its constancy?

10.1 Perturbations of inconstant shape: the missing baroclinic wave

A major forecast problem is concerned with forecasting development; intensification of more-or-less observable existing systems. While exponentially amplifying waves are a useful description of some aspects of wave generation, they are not the whole story, and there are some paradoxical cases. For example, consider the Eady system, in which the baroclinity is independent of height, with no variation of inertial density, and constant stratification, but with β not necessarily zero, as illustrated in Figure 10.1. With β zero there is a short-wave cut-off to instability at $k = 2.4$ nearly, and one pair of non-amplifying waves with steering levels above and below the middle level for shorter waves. But with $\beta = +0.01$ there is no short-wave cut-off, but only a pair of short waves with steering level near the lower boundary, one amplifying weakly, the other diminishing weakly. The wave that had its steering level near the upper boundary has vanished. Conversely, if β were small and negative, then the lower-level wave would disappear. We argue that these disappearing waves are part of a continuous spectrum of solutions that occasionally attain the property of being of constant shape. It could be argued that to make the effective (i.e. the non-dimensional) value of β change sign we have to increase the shear through

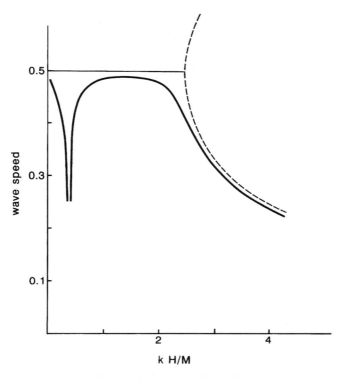

Figure 10.1 Variation of phase velocity with wavenumber for Eady's solution with $\beta = 0$ and $\beta = 0.01$. For $\beta = 0$ shown as the dashed line, there is only one pair of amplifying/diminishing waves up to wavenumber 2.4, and a pair of non-amplifying waves with steering levels above and below the middle level for shorter waves. For $\beta = 0.01$, shown as the thick line, there is an amplifying/diminishing pair for all wavelengths with steering levels near the lower boundary, consistent with retardation due to the β-effect associated with Rossby waves. We argue that these disappearing waves are part of a continuous spectrum of solutions which occasionally attain the property of being of constant shape.

$+\infty$ to $-\infty$ and therefore perhaps one should not expect physical continuity, but we can think of other ways of changing the apparent value of β, by changing the horizontal variation of the mean wind for example, so I think we should be worried.

Moreover the simple parcel theory of slantwise convection of Section 9.5 equates the energy of the perturbation with that released by rearrangement of the original baroclinic distribution of potential density which is not changed by the value of β. I believe that this rearrangement is the essence of the generation of baroclinic waves, and that the vorticity equation is a secondary constraint that also has to be satisfied, in much the same way as we view vertical over-turning. We do not expect the equations of motion to prevent warm air rising relative to cold in ordinary convection, so why should we expect such a constraint in slantwise convection?

Following this line of argument, the vorticity equation becomes merely a diagnostic for the scale, and stability criteria as discussed in Chapter 9, are misleading. However the case where the fluid is very deep is not consistent with this view. When β is positive, we find the solution of Charney (1947) where amplifying waves exist for almost all wavelengths, but for $\beta = 0$ the only separable solution to equation 9.31 is

$$\psi = a \exp \mathrm{i}k(x - ct) \exp -kz/M$$

where

$$c = AM/k \tag{10.1}$$

The phase velocity c is real showing that there are no amplifying waves, in spite of the system having the same store of available potential energy as when β was positive. Is the parcel description of slantwise convection perhaps too simplified?

10.2 A complete set of solutions

So far, we have concentrated attention on waves of constant shape. For a given x-wavelength, there are not many of them; in the Eady problem only two and in the Charney limit of it only one. This is certainly not enough to express variation with height of an arbitrary initial flow, so there must be many others to make up a complete set.

If we set up a two-point centred finite difference grid for the differential equation 9.31 with boundary conditions 9.32, then with $U'' = \beta = 0$ the complete set of solutions is given by a set of steadily propagating waves that have steering levels at each interior grid point, together with the amplifying/diminishing pair of Eady. While this comprises a complete set of solutions, it is not very satisfying. The solutions are model-dependent, and have a temperature discontinuity, associated with a discontinuous change in the vertical gradient of stream function, at their steering levels. There is a more satisfying fundamental set.

Consider the general solution to the linear forecasting problem, given a vanishingly small, but otherwise perfectly arbitrary, initial perturbation to the Eady system. No generality is lost if we concentrate on motion which is independent of the y-coordinate, and deal with one x-wavelength at a time, but we must leave the temporal variation arbitrary. The defining equation is, with $U = Az$,

$$(\partial/\partial t + \mathrm{i}kAz)(M^2\psi'' - k^2\psi) = 0 \tag{10.2}$$

which has the first integral

$$M^2\psi'' - k^2\psi = a \exp -\mathrm{i}kAzt \tag{10.3}$$

where a should be an arbitrary function of z. Astonishingly, however, we can take a as a constant and integrate again to give

$$\psi = a(1 + M^2 A^2 t^2)^{-1} \exp ik(x - Azt) \tag{10.4}$$

These functions are not of constant shape, but get blown over by the shear. They depend for their usefulness on having an undetermined time origin. Thus we can think of them as starting at some non-zero time t_0. Then equation 10.4 represents a family of functions that oscillate in the vertical like $\exp ikAzt_0$. At the initial time they, with appropriate coefficients $a(t_0)$, form a complete set for Fourier synthesis of an arbitrary initial function. It is likely that both signs of the phase will be needed, so there will be some functions with positive t_0, and some with negative t_0. According to equation 10.4, these latter will increase in amplitude up until the time $t = -t_0$, at which their phase lines become vertical, and then they begin to decrease in amplitude again.

10.3 A complete solution

As it stands equation 10.4 does not satisfy the boundary condition; $w = 0$ at a rigid lid, or

$$\partial^2 \psi / \partial z \partial t - ikA \psi = 0 \quad \text{at } z = 0 \tag{10.5}$$

but any term $F(t) \exp -kz/M$ can be added, for this satisfies equation 10.2 for any functional form of $F(t)$ and the sum of it and equation 10.4 satisfy the condition equation 10.5 at the lower boundary if

$$\frac{dF}{dt} + iAM F = -\frac{2iAMk^2}{(1 + M^2 A^2 t^2)^2} \tag{10.6}$$

The temporal boundary condition $F = 0$ at $t = t_0$ leaves the completeness of equation 10.4 unchanged, so that the solution at $t = t_0$ can then be represented by the Fourier sum of functions like those in equation 10.4. The solution to equation 10.6 is an exponential integral, but it is simple to integrate the equation numerically. Because the timescale is $1/AM$ on both sides, the term on the RHS behaves nearly as a temporal point source centred at $t = 0$ where the solution grows like

$$\partial/\partial t (\log F) = 2M^2 A^2 t / (1 + M^2 A^2 t^2)^2 \tag{10.7}$$

Thus for AMt large and negative F increases like $(1 + A^2 M^2 t^2)^{-2}$, then F jumps rather quickly, as can be seen in equation 10.7, to a finite multiple of $\exp -iAMt$ as we pass through $t = 0$.

Figure 10.2 shows a numerical solution for the amplitude of the stream function at $z = 0$; $F + 1/(1 + A^2 M^2 t^2)$. What we see is the rapid growth of a

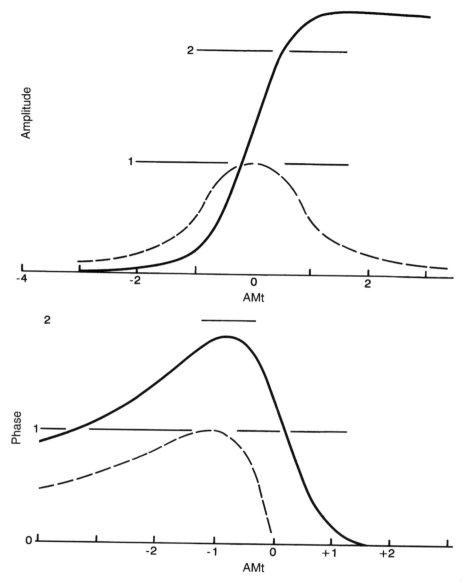

Figure 10.2 Variation with non-dimensional time AMt of the amplitude and phase of two parts of an inconstant shape wave. Dashed line is ψ_1; the part that is advected with the local flow and which appears throughout the fluid at the same time. Continuous line is ψ_2; the part that decreases exponentially away from the boundary, and is the same shape as the non-amplifying wave of the classical analysis.

system that has the shape of the neutrally stable solution. Maybe this is the missing baroclinic wave.

This boundary wave of constant shape (ψ_2) grows in association with the inconstant-shape wave (ψ_1) whose tilting troughlines appear throughout the whole depth of the fluid. Perhaps we can picture the inconstant wave as channelling energy from the bulk of the fluid towards the boundary wave?

In some ways these waves mimic the life history of the constant-shape waves, which grow in amplitude, become non-linear, and eventually die. Or do they, for may not the non-linear remnant of the original wave become the inconstant shape precursor for the next? We remind ourselves that constant-shape amplifying waves have the property not shared by the inconstant-shape solutions of necessarily growing to finite amplitude thus demanding non-linearity, and the consequent modification of the original system.

10.4 Rate of amplification of the inconstant wave

Equation 10.7 defines a growth rate. This is small both for t small and t large, with a maximum value of MA at $MAt = 1$. This contrasts with the solutions for the constant-shape wave whose thermodynamic maximum amplification rate was $MA/2$, with amplification rate of $0.3\,MA$ for realistic boundary conditions, and which did not amplify at all in the system without an upper lid that we study here.

That the inconstant wave appears to grow faster than the parcel theory gives as an upper bound is suspicious. After all, the parcel theory was a sort of optimum system where particles were supposed to arrive at the best position for energy release, independent of constraint. However, in contrast to waves of constant shape, the amplitude of the stream function is not the best guide to the energy of the more general wave. Thus we notice that when t_0 is large, then at the initial time $t = 0$, ψ_1 varies rapidly with height, and that since $\partial\psi/\partial z$ represents the temperature perturbation, the wave potential energy must be large compared with the wave kinetic energy. As time progresses the vertical variation, and therefore the wave potential energy decreases, eventually to zero, as the amplitude of ψ_1 increases.

The total energy of the perturbation is the sum of the potential energy P and the kinetic energy K, given by

$$P = \frac{1}{2}\,\phi^2 = k^2 A^2 t^2 \mid \psi_1 \mid^2 \text{ and } K = \frac{1}{2}\,v^2 = k^2 \mid \psi_1 \mid^2 \qquad (10.8)$$

Thus the wave starts off with comparatively larger potential than kinetic energy, which it converts into kinetic energy by the time the wave reaches maximum amplitude. Only the remainder of the energy is extracted from the mean flow, as envisaged by the parcel theory.

Equation 10.8 suggests that a better representation of the total wave energy might be

$$P + K = k^2(1 + M^2 A^2 t^2) \mid \psi_1 \mid^2 = a^2 k^2/(1 + M^2 A^2 t^2) \qquad (10.9)$$

and a better representation of the amplitude of the energy of the system would then be $(1 + M^2 A^2 t^2)^{-1/2}$, whose exponentialised growth rate is $-MAt/(1 + M^2 A^2 t^2)$ This has maximum value $0.5MA$ at time such that $MAt = -1$, curiously the maximum value predicted by the parcel theory, and significantly larger than the maximum growth rate (0.31) deduced from the dynamic theory taking into account the constraint of a horizontal lid.

10.5 General baroclinic waves with two lids

In a more realistic case there are two boundaries separated by a height difference H. In this case it is convenient to select a coordinate system in which $-H/2 < z < H/2$. The function ψ_1 remains the same, but we can now add another constant-shape wave, say of the form $P(t) \exp +kz/M$ to help satisfy the new (upper) boundary condition. It is convenient to take ψ_2 of the form

$$\psi_2 = a\left(F \cosh \frac{k(z + H/2)}{M} + G \cosh \frac{k(z - H/2)}{M}\right) \bigg/ \sinh \frac{kH}{M} \quad (10.10)$$

We take the unit of time to be $1/AM$, and the unit of horizontal distance to be H/M. This device allows us to write AMt as t and kH/M as k so long as we recognise that these variables must be measured in these intrinsic units. This step avoids the clumsy mathematical device of non-dimensionalising the equations, then relabeling all the variables. Either way the two boundary conditions become

$$\left(\frac{\partial}{\partial t} - i\frac{k}{2}\right)G + i\left(\frac{F}{\sinh k} + \frac{G}{\tanh k}\right) + \frac{2i \exp +ikt/2}{(1 + t^2)^2} = 0$$

and

$$\left(\frac{\partial}{\partial t} + i\frac{k}{2}\right)F - i\left(\frac{F}{\tanh k} + \frac{G}{\sinh k}\right) - \frac{2i \exp -ikt/2}{(1 + t^2)^2} = 0 \quad (10.11)$$

The differential equations 10.11 satisfied by F and G are such that G is the complex conjugate of F. This is the same vertical symmetry that makes for the delightful simplicity of the Eady system, and the astonishingly simple representation of the highly truncated non-linear system of Lorenz, equation 13.17.

Again the last term behaves much like a temporal point source and serves to generate a wave as t passes through zero. However, it is now an exponentially amplifying wave rather than the neutral one of the previous section that is generated. The temporal boundary condition is again $F = G = 0$ at the initial time $t = t_0$. The repercussions are perhaps not immediately obvious but Figure 10.3 shows the amplitude of the surface stream function for some of the solutions.

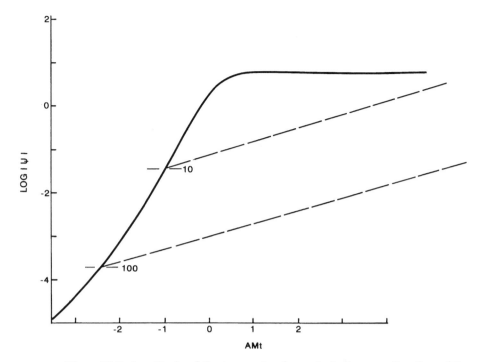

Figure 10.3 Amplitude of the two parts of a perturbation as a function of time. We assume that the wave of constant shape increases in amplitude like exp $0.3\,AMt$ and that it is of similar amplitude to the inconstant wave at the initiation of the perturbation. The critical factor governing which species of wave will finally emerge is the ratio of the initial pertubation to that of the wave when it becomes non-linear. Though the inconstant wave grows much faster, it has not long to live. Illustrated is the reasonable value $AMt_0 = 1$.

Remarkably their evolution is rather similar to that of the boundary wave, at least up until $t = 0$. Thus when t is large and negative, growth is like t^{-4} rather than exponential. For $t < 0$ the inconstant growth dominates, becoming exhausted and being replaced by the exponential near $t = 0$. As before, the inconstant wave has different evolution for the kinetic and potential energy, so let us concentrate attention on the amplitude of F which essentially represents the kinetic energy of mostly the lower wave, i.e. the intensity of the surface low.

New waves are not initiated at infinitesimal amplitude, but develop in the context of some finite-amplitude flow. To compare the amplification of the two regimes of growth, we need to know where on the growth curve of the inconstant wave that we start from. The temporal initial condition selects inconstant shape waves with vertical structure like exp $ikAzt_0$. Since these are combined in some way with functions like exp $-kz/M$ at this initial time, we might suppose the two height scales to be comparable, i.e.

$$kAt_0 \sim k/M \qquad\qquad (10.12)$$

This gives AMt_0 of about one. This number is critical, for though the growth rate of the inconstant wave is some six times as large as that of the exponential wave, as we see from Figure 10.3, it amplifies only until a value of AMt of about 0.5.

Taking $AMt_0 = -1$, and using the solutions shown in Figure 10.3, we see that the inconstant wave would amplify by a factor of about ten during this time, then stop. In contrast, the exponentially growing wave would amplify by a factor of only $\exp(0.30 \times 2.5) \simeq 2$ during the same time, but it would keep on growing afterwards. Thus if the initial perturbation is bigger than $1/10$ of the final perturbation, the inconstant wave will win. If not, the exponential wave will win, though we will have to wait until $AMt > 7$ before it does. If the exponential wave had twice the amplitude of the inconstant wave, then it would take only until $AMt > 3$, before the exponential wave became of greater amplitude.

If AMt_0 is larger, meaning that the vertical distribution of the initiation is more oscillatory, then the inconstant wave becomes dominant very much more rapidly. Frank Ludlam (private communication) often spoke of troughs in the upper troposphere starting in the jet off the eastern coast of America and triggering the development of weather systems when they got to Europe. I tried to convince him that they were carried by the group velocity of stationary Rossby waves, which gives the right speed, but neither of us were convinced. Perhaps we have here a more rational explanation?

10.6 Predictability

This is one of those topics apon which I have no proper basis for comment. I do not have a feel for the abrupt difference between deterministic, and undeterministic systems that seems to pervade much of the literature. It seems to me that, whatever system I forecast, there is a given time ahead at which I cannot decently expect to get the phase of an oscillation correct, and if the state of the system depends on the mutual phases of several different components then the chances of good predictability is small. Similarly, pictures of indeterminate (chaotic) systems seem to show them going round the same old circuit many times before 'unexpectedly' going off into another orbit. Apart from the transitions, they seem particularly easy to predict; maybe their unpredictability is all squeezed up into a localised part of parameter space?

Some systems are more predictable on a longer timescale. Sunshine poured into the boundary layer will produce fairly predictable heating of the boundary layer, but on the shorter timescale it is difficult to predict where the next convective element will start. This is a bit like predicting that a drunken man carrying a tray of drinks will eventually spill some on the floor, but which one goes first is more debateable.

Finally, I am perplexed by the proposition that the successive digits of π are not related to each other, in the sense that they are indistinguishable from a random selection of digits. Unless, that is you know that they are the digits of π, in which case they are perfectly predictable! Maybe it is the way we try to do our prediction, all this step by step integration for example, that is at the root of unpredictability rather than an intrinsic property of the system itself. It always seems to be the simpler and the less dissipative systems that are more unpredictable, the more forced complicated damped ones that are more predictable.

Chapter 11

Motion in a barotropic atmosphere

11.1 The barotropic quasi-geostrophic vorticity equation

For rather long waves and when the horizontal gradients of temperature are suitably small, the effect of the slantwise nature of the motion can be ignored and the quasi-geostrophic vorticity equation for motion of small amplitude about a constant zonal flow U and constant static stability B, and ignoring the variation of inertial density with height, becomes

$$\left(\frac{\partial}{\partial t} + U\frac{\partial}{\partial x}\right)\left(\frac{\partial^2 \psi}{\partial x^2} + \frac{\partial^2 \psi}{\partial y^2} + M^2\frac{\partial^2 \psi}{\partial z^2}\right) + \beta\frac{\partial \psi}{\partial x} = 0 \tag{11.1}$$

Solutions of equation 11.1 can be written

$$\psi \propto \exp ik\,(x - ct)\,\exp \pm i\nu z$$

so long as

$$c = U - \frac{\beta}{(k^2 + M^2\nu^2)} \tag{11.2}$$

The quantity c is the horizontal component of the phase velocity of the wave. The group velocity can be defined in terms of the state of the system a 'long time' later and therefore in terms of the stationary phase of the frequency $kx - kct + \nu z$ with respect to variations in k and ν which gives

$$(\delta x/\delta t)_{gr} = \partial(ck)/\partial k$$

and

$$(\delta z/\delta t)_{gr} = \partial(kc)/\partial \nu \tag{11.3}$$

We notice from equation 11.2 that the waves move more slowly from west to east, than the air, with the effect most marked for longer waves. Some waves will have $c = 0$ and therefore be stationary relative to the ground. From equation 11.3, we notice that such stationary waves will, in general, have non-zero group velocity. This means that we should expect to see a patch of stationary waves spread out from the origin of the disturbance, at the group velocity, as time proceeds.

All values of v may not be allowed. If the fluid is bounded by rigid horizontal lids at $z = 0$ and $z = H$, then the boundary condition $w = 0$ demands $\partial\psi/\partial z = 0$, so ψ must be proportional to $\cos vz$ where $vH = n\pi$ with n an integer. Such waves are free waves, as distinct from waves forced by some external influence. We wonder what physical processes might generate them. We notice that the motion of the air in the waves is not, in general horizontal; the fluid is barotropic but the waves are not.

11.2 Stationary waves

We are particularly interested in those waves that are stationary relative to the ground, for they may be generated by topographic features, and have climatalogical influence. For these waves $c = 0$, and the stream function is given by

$$\psi = a \exp \mathrm{i}(kx + vz) + a' \exp \mathrm{i}(kx - vz)$$

where

$$k^2 + M^2 v^2 = \beta/U \tag{11.4}$$

At a smooth lower boundary whose (small) height is given by the real part of $h_0 \exp \mathrm{i}kx$, $w = U\, \partial h/\partial x$ or in terms of the stream function

$$w = -\mathrm{i}kU \frac{f}{N^2} \left(\frac{\partial\psi}{\partial z} \right) = \mathrm{i}kUh \quad \text{at } z = 0$$

or

$$a - a' = \mathrm{i}\frac{N^2}{fv}h_0 \tag{11.5}$$

One more boundary condition is needed to determine both a and a', and this is likely to be determined by the nature of the upper limit to the system.

11.3 Lids lead to resonance

> had we but world enough, and time

If the upper boundary is a rigid lid the additional boundary condition is $w = 0$ at $z = H$. Using equation 11.5 gives $a\mathrm{e}^{\mathrm{i}vH} = a'\mathrm{e}^{\mathrm{i}vH}$ and finally

$$\psi = -\frac{N^2 h_0}{f\nu} \frac{\cos \nu(z-H)}{\sin \nu H} \exp ikx \qquad (11.6)$$

The response is unbounded for $\nu = 0$, π/H, $2\pi/H \dots$ which picks out the free waves. This unboundedness is to some extent a consequence of the steady-state assumption, for that gave unlimited time during which the flow might indeed acquire unlimited amplitude, as a simple pendulum forced at its resonance frequency from a state of rest does, at least while the mechanics remains linear.

There are two simple ways of avoiding this unboundedness in the steady state. We can introduce dissipation to absorb the energy, or make the channel infinitely long or high, thereby allowing the energy to disperse. In many respects these have similar repercussions, but dispersion and dissipation are not always replacements for each other. In this case with a closed channel they serve to make the fluid forget the energy source before it comes round to the energy source again.

11.4 Surface friction

Consider a surface friction layer in which an additional vertical velocity proportional to the relative vorticity is generated as in Section 8.4

$$\Delta w = v_* \zeta / f \qquad (11.7)$$

where the stress $\tau = \rho v_* \mathbf{v}$ and v_* is a constant 'friction velocity'. Adding this contribution to the vertical velocity induced by the mountain in equation 11.5 gives

$$\psi = -\frac{(N^2 h_0/f) \exp ikx}{\nu \sin \nu H + i(N^2 k v_*/f\nu) \cos \nu H} \qquad (11.8)$$

The motion is now bounded for all (real) values of ν. Solutions of equation 11.6 were unbounded for $\nu H \ll 1$, in which limit equation 11.8 becomes

$$\psi = \frac{fU h_0 \exp ikx}{H(k^2 U - \beta) - ikv_*} \qquad (11.9)$$

Now the amplitude remains finite, even at the resonant wavenumber $k = (\beta/U)^{1/2}$. The amplification is still quite large, with HkU/v_* of about 10 for a typical value $v_* \simeq 1\,\mathrm{cm\,s^{-1}}$. This is a nice illustration of the modeller's dilemma. He can choose the value of the parameter in the sub-gridscale process that gives the best fit to the data, and argue that this is a justifiable way of observing this rather obscure parameter that he has just invented. In this case the quantity v_* has to represent the frictional influence resulting from all scales of motion which are smaller than the grid length in the model, which we could not hope to measure directly. Or should he infer an independent value that might give a large discrepency between theory and observation but leave the way open to further investigation. There is some evidence that rather large values of friction are needed to get the best fit. Can other dissipative mechanisms be identified?

11.5 Friction at an upper lid

Suppose that energy was propagated away from the lower layers to be dissipated higher up, perhaps due to shearing turbulence in the jet stream. To mimic this we could invent an explicit sink of energy at an upper level. Suppose, for example that there was an impenetrable, but frictional boundary at $z = H$. The boundary condition is then the reverse of that in Section 8.4; $w = -v^*\zeta/f$ at $z = H$ and v^* is the appropriate value of the friction velocity for upper-lid friction. The solution is

$$\psi = \frac{N^2 h_0}{f v} \frac{(\cos\, v(z - H) + iC \sin v(z - H)))}{(\sin vH + iC \cos vH)} \exp ikx,$$

where

$$C = \frac{v^* k N^2}{U v f^2} \tag{11.10}$$

Notice that the phase of this solution changes with height, showing that energy is propagated in the vertical direction. Indeed only with $C = 1$ is there no downward propagation.

When the upper friction is very large the solution reduces to

$$\psi = \frac{N^2 h_0}{f v} \frac{\sin v(z - H)}{\cos vH} \exp ikx \tag{11.11}$$

in which ψ vanishes at the upper surface, showing that the friction there has been so strong as to prevent any motion at all, and incidentally, to make the lid a perfect reflector of energy. Unimpeded transmission of energy upwards, represented by the special value $C = 1$, needs a carefully matched impedance. What happens when we have that?

11.6 Very deep atmosphere

Consider the case of a very deep atmosphere. Both parts of the solution 11.4 are bounded as $z \to \infty$, so there is nothing to choose between them on that account. We notice that, as in the solutions with friction in the upper atmosphere, the phase changes with height. Thus relative to a particle travelling through the wave, the phase in the first term of equation 11.4 propagates upwards away from the ground, which is the source of the disturbance,

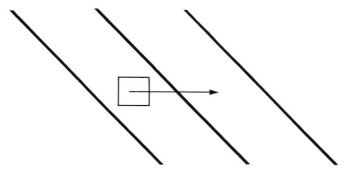

Figure 11.1 A parcel of air flowing through phase lines oriented downwind sees phase velocity directed upwards, but the vertical components of phase and group velocity are in opposite sense so flow near the uneven ground tends to have the opposite orientation. We note that this is consistent with a drag, rather than a pull on orographic irregularities through the Bernoulli pressure.

shown in Figure 11.1. At first sight this seems a good thing. On further study we find that the phase and group velocities are in the opposite sense, so that the energy propagation in the first term is *towards* the source. Thus if we demand that no energy is returned we must set $a' = 0$. This is a simple example of the so-called radiation condition, sometimes expressed as the requirement for propagation of energy away from the disturbance, though it is better thought of as the condition for no reflected energy, since all solutions with $a > a'$ radiate energy upwards. The solution satisfying this perfect radiation condition is

$$\psi = i\frac{N^2}{f\nu}h_0 \exp i(kx + \nu z) \tag{11.12}$$

which is finite except possibly for the special case $\nu = 0$. For all solutions satisfying a radiation condition the energy is transmitted upwards and does not return. How is it lost, we wonder?

11.7 Upward propagation of energy

A very simple initial value problem is illuminating. Consider the time development of motion over a simple sinusoidal surface. Suppose that initially, there was no flow, but that it starts up from rest at $t = 0$.

The temporal boundary condition demands $\psi = 0$ at $t = 0$ and

$$w = -\frac{f}{N^2}\left(\frac{\partial}{\partial t} + U\frac{\partial}{\partial x}\right)\frac{\partial \psi}{\partial z} = ikh_0 U \quad \text{at } z = 0 \text{ for } t > 0 \tag{11.13}$$

It is convenient to express the solution as the sum of two terms; one independent of time and satisfying the boundary condition at the ground which is also independent of time at least for $t > 0$, and a linear sum of free travelling waves. Since these have zero vertical velocity at the ground they do not affect the boundary condition there, but they can be chosen to satisfy the condition $\psi = 0$ at $t = 0$. The stationary contribution can be written

$$\psi = a_0 \sin \nu_0 z \, \exp ikx$$

where

$$a_0 = -\frac{N^2 h_0}{\nu_0 f} \text{ and } k^2 + M^2 \nu_0^2 = \frac{\beta}{U} \tag{11.14}$$

The free wave is

$$\psi = a(\nu) \cos \nu z \, \cos ik(x - ct)$$

where

$$c = U - \frac{\beta}{(k^2 + M^2 \nu^2)} \tag{11.15}$$

It remains to choose that particular sum of travelling waves that just cancels the stationary wave at the initial time. What we need is a sum of cosines that is equal to a sine. Unlikely though this might seem in view of the different behaviour of $\sin z$ and $\cos z$ near $z = 0$, the sum is in this case fairly simple. Integrating over all vertical wavenumbers ν keeping the horizontal wavenumber k fixed we find

$$a(\nu) = \frac{\nu_0}{\pi(\nu_0^2 - \nu^2)} \tag{11.16}$$

Integration along the real ν-axis from $-\infty$ to $+\infty$, with an indentation towards the positive-imaginary side to circumnavigate the poles at $\nu = \pm\nu_0$ as shown in Figure 11.2 gives the desired finite contribution. Off we go into the complex plane again, is it for the same reason as before or another?

We can begin to visualise the real integral implied by Equation 11.16, with a lot of $\cos \nu_0^+ z$ taken away from a lot of $\cos \nu_0^- z$ to leave a lot multiplied by the small difference between the two cosines which would be like the sine of the average, but that hardly seems to aid physical interpretation. It all seems a bit elaborate, and this wandering off into the complex plane demanding of unlikely mathematical techniques, but let us bear with it for a while to see what it has to offer in the way of enlightenment.

The solution cannot be reduced to closed form in terms of simple recognisable functions except at $t = 0$. Indeed for large t the term in the integrand $\exp ikct$ is very badly behaved in the sense that it oscillates so violently with respect to ν as to be largely cancellatory. But this turns out to be some salvation for we can now seek those places, perhaps in the complex plane, where the term locally ceases to oscillate, and where the contribution to the integral may be large. This point of 'stationary phase' in the ν-plane is found where

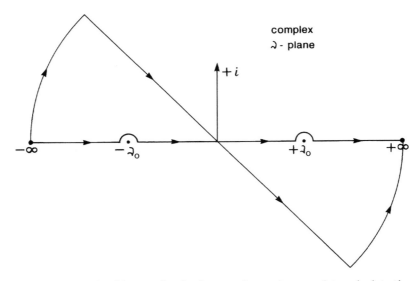

Figure 11.2 Path of integration in the complex v-plane used to calculate the integrals involved in equation 11.16.

$\partial(kct)/\partial v = 0$. This is near $v = 0$ for at least some coordinate values. Expanding about this point of stationary phase gives

$$c = c_0 + v^2 c_2 + \ldots$$

where

$$c_0 = U - \frac{\beta}{k^2} \qquad c_2 = \frac{\beta M^2}{k^4} \qquad (11.17)$$

The theory of residues says that we can take any convenient path in the complex plane, between the limits of integration, taking due account of any poles that are circumnavigated. If t is large enough, the term in v^2 dominates. Since c_2 is positive the path of *steepest descents* follows $v = (1-i)s$ where s is real as in Figure 11.2. This part of the integral is easily seen to be of order $t^{-1/2}$. Thus for t suitably large, the path from $v = -\infty$ to $+\infty$ can be diverted to the quadrant $v = (-1+i)\infty$ then along the steepest descent to $v = (1-i)\infty$, thence to $v = +\infty$. Taking account of the deviation to the negative-imaginary side of the pole at $+v_0$ we find the contribution to the integral round that is just $i\cos v_0 z$, so the final solution for large time is

$$\psi = -ia \exp i(kx + v_0 z) + O(t^{-1/2}) \qquad (11.18)$$

This then is the solution generated through time evolution. It is stationary and it is *the* solution having zero downward energy propagation.

In contrast, a lid, and stiff friction both give the same amount of downward as upward propagation of energy, independent of where the reflecting layer happens to be. Reality is more complicated and somewhere in between.

The concept of energy propagation introduces that of group velocity. Indeed had we examined the phase $(vz - kct)$ for stationarity, instead of just the kct-part, then the vertical component of the group velocity defined by equation 11.3, would have appeared immediately.

We are astonished by the excursion into the complex plane required by such analysis. While we can visualise a velocity profile $U(z)$ for real values of z, we find it difficult to imagine it being explored in the complex plane. The idea of the wind at a complex point in space, evidently important in this development, is somewhat daunting. Maybe the complex plane is not such an inhospitable place as textbooks on mathematics lead one to suppose; maybe a point in the complex plane is very much dependent on what it can 'see' of the real variable?

Alternative formulations of the steady state problem include the addition of Newtonian damping in which $\partial/\partial t$ is replaced by $\partial/\partial t + \alpha$. This serves to bring the motion back towards an undisturbed state, with time constant $1/\alpha$, and without increasing the order of the equations, as the addition of molecular viscosity would for example. This device is popular with simulator-modellers who want to persuade their model that they know better than their model what the answer should be, by guiding the results towards reality using a Newtonian drift towards that state.

We can let the height of the mountain grow exponentially with time. At least in several simple cases each of these have the effect of declaring a direction for circumnavigation of the singularities on the real axis, and agree with the dispersive theory on the final solution selected.

Perhaps we can avoid this excursion into the complex plane by using the sort of ray tracing techniques that are so powerful in geometric optics; using the local group velocity to transmit amplitudes and wavelengths across the fluid to reconstruct the flow field further away?

11.8 Propagation in the real atmosphere

In practice, the zonal wind is not independent of height, but it may be so slowly varying that we can apply a theory much like this at each level. The condition for propagation then depends on the vertical profile of v_0. Propagation is most likely for small k, i.e. for long waves, and is easiest where U is small, as long as it is positive. The static stability B which we might have supposed governed propagation, only serves to modify the vertical length scale. Thus we might suppose that the average value of U within about $1/v_0$ of the tropopause governs the propagation of long-wave energy into the upper atmosphere. During the summer there is a deep band of easterlies (negative U) in the lower stratosphere, which prevents energy getting through. During the winter, the jet maximum reflects all but zonal wavenumbers 1 and 2 which then dominate the stratospheric flow.

It is more complicated than that though, for there is also a meridional variation in the wind. Not only does the tropopause wind vary with latitude, but also the wind near the surface that generates the vertical velocity at the ground. More general theory shows that some components of the wave energy tend to be deflected in the meridional plane, towards the poles, where it contributes towards the breakdown of the polar stratospheric jet. Consider a simple example of such latitudinal dispersion.

11.9 Lateral dispersion

Baroclinic activity is most intense in the regions of strong horizontal temperature gradient, and therefore tends to produce trains of positive and negative vorticity anomaly concentrated in middle latitudes. Eady (not neccesarily first, or exclusively) thought that this might disperse because of the β-effect. Suppose for example that at $t = 0$ the baroclinic activity has created a train of barotropic lows and highs confined to a latitude belt, giving a stream function of the form

$$\psi = A \cos kx \exp -(y^2/a^2) \tag{11.19}$$

The y-variation can be decomposed in terms of propagating Rossby waves to give

$$\psi = \frac{aA}{2\pi^{1/2}} \int_{-\infty}^{+\infty} d\mu \, e^{-a^2\mu^2/4} \, \exp ik(x-ct) \, \exp i\mu y$$

where

$$c = U - \frac{\beta}{(k^2 + \mu^2)} \tag{11.20}$$

For large time the stationary phase is for $\mu = \pm k^3 y/2\beta t$, positive for positive y, negative for negative y, and we find the asymptotic solution

$$\psi = \frac{1}{2} aA \left(\frac{k^3}{\beta t}\right)^{1/2} \exp i \left(kx - kc_0 t + \frac{k^3 y^2}{2\beta t} + \frac{\pi}{4}\right) \tag{11.21}$$

A numerical solution, illustrated in Figure 11.3 shows the streamlines bowing out in such a parabolic shape in the horizontal plane, as the vorticity disperses. This is in the sense for energy to be propagated away from the source region, but it is also in the sense to give convergence of zonal momentum into the source region, through the horizontal Reynold's stress term $\overline{u'v'}$. Is this not curious?

From the numerical solution we see the amplitude spreading out through $t = 5$ to $t = 16$ with the zone of sloping phase lines spreading out with it, leaving

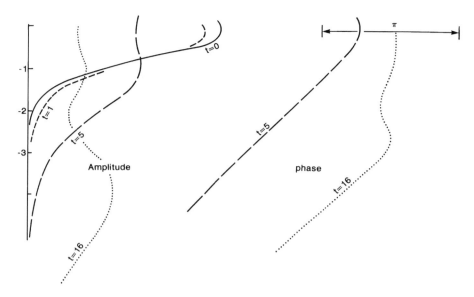

Figure 11.3 Dispersion of energy away from the latitude of origin gives troughlines a parabolic shape. This shows that dispersion of energy away from a zone implies a Reynolds' stress directed towards that zone. We see that the disturbance spreads out as time progresses leaving a region of waves long in the *y*-direction behind.

behind a central region of more uniform amplitude and less phase change. I suppose that we can see this in an appropriate expansion of the Fourier integral. The maintenance of mean surface winds against the dissipative effect of surface friction by the transfer of the zonal component of momentum against the mean gradient of the zonal component of momentum is a puzzling feature of the general circulation of the atmosphere, discussed further in Section 14.5, and this dispersion is a suggestion for a mechanism. Curiously this idea of momentum transfer into the baroclinic zones, was suggested by Roland Davies as being heuristically plausible from first principles, in about 1976, but I can't find a published reference so maybe it was another one of those personal communications.

11.10 Longitudinal horizontal dispersion

Baroclinic activity varies with longitude, as well as latitude. Thus we see baroclinic waves amplifying in the baroclinic region off the east coast of North America, particularly in the winter, or directly upstream of blocking anticyclones. Over the eastern Atlantic shorter waves associated with enhanced precipitation develop in the frontal regions of the mature longer waves. As with the previous example the actual shape of the initial concentration seems unimportant, and we gain insight with a disturbance with an unbounded y-wavelength, and put

$$\psi = \int_{-\infty}^{+\infty} A(k) \exp ik(x - ct) \, dk \text{ with } c = U - \beta/k^2 \qquad (11.22)$$

It is convenient to use the Lagrangian coordinate $x' = x - Ut$, but here we choose to treat the special case $U = 0$ which achieves the same end. When the initial disturbance is isolated in space, A is a slowly varying function of k and we find points of stationary phase where $k^2 = \beta t/x$. We can expand about this point so long as $| \beta xt |$ is large. Solutions for positive and negative values of x behave rather differently. For x positive, contributions from the $+$ve and -ve values of k give

$$\psi = \pi^{1/2} A(0) \left(\frac{\beta t}{x^3}\right)^{1/4} \cos \left(2(\beta xt)^{1/2} + \pi/4\right) \qquad (11.23)$$

For negative values of x there is only one possible point of stationary phase, which gives

$$\psi = \pi^{1/2} A(0) \left(\frac{-\beta t}{x^3}\right)^{1/4} \exp -2(-\beta xt)^{1/2} \qquad (11.24)$$

It is surprising at first sight that the amplitude dies away rapidly in the negative x direction, which is in the direction that the waves propagate, and extend comparatively far towards positive x, where the waves do not. All a bit puzzling but a consequence of the group and phase velocities being of opposite sign relative to the air, of course. Figure 11.4 shows the Fourier integral for a pulse initially isolated in the x-direction calculated numerically. We see the phase propagate towards negative x while the amplitude propagates towards positive x. Figure 11.5 shows the same data on a time-distance plot. Even such simple two-dimensional, linear, experiments are not easy to visualise.

We can check some aspects of this behaviour analytically, by introducing an imaginary wavelength that serves to describe some aspects of the two ends of the wave packet. Thus we see that one possible solution to the vorticity equation 11.1 is

$$\psi = \exp(\alpha x - \beta t/\alpha) \qquad (11.25)$$

For x large and negative α must be positive, and if it is also large, to mimic the steep wave packet, then the solution 11.25 will decrease with time, but slowly, similar to equation 11.24.

Similarly, for x large and positive, α must be negative, and solution 11.25 amplifies with time, with the components that decrease more slowly with distance amplifying more rapidly in time.

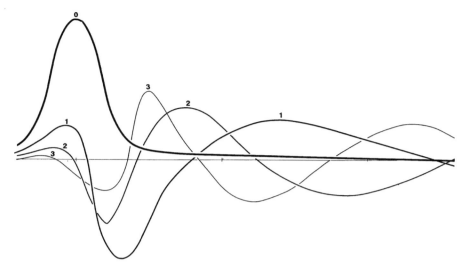

Figure 11.4 A pulse at $t = 0$ is relatively isolated in the x-direction and spreads out as a wave packet. We see that the phase propagates towards negative x while the amplitude propagates towards positive x.

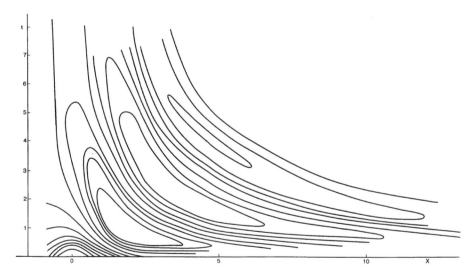

Figure 11.5 Shows the same information but on a time–distance plot. Even such simple two-dimensional, linear, physics is not easy to visualise.

11.10.1 A curious similarity solution

The Fourier integral 11.22 is similar to one of the formulae for Bessel functions which suggests another technique for examining solutions. If we put $\psi = 1/(a-\mathrm{i}x)$ at $t = 0$, intending to examine the real part only, then $A(k) = A_0 \exp -ak$. This trick gives

$$\psi = A_0 \int_o^\infty dk \, \exp\left((ix - a)k + i\beta t/k\right) \qquad (11.26)$$

and by changing the variable of integration to $\theta = k/\beta t$, we find that the solution must be of the form $\psi = tf(\chi)$ where $\chi = (x + ia)\,\beta t$. This is one of the self-similar solutions usually met in the description of diffusion near point sources, unusually for dispersion as here. Is that not curious? The shape of the solution is determined by χ, which transforms roughly into a space coordinate through the multiplier βt whereas the amplitude of the solution is determined through the multiplier t. Substituting back into the original dispersion equation $\partial^2 \psi/\partial x \partial t + \beta \psi = 0$ we find that f must satisfy

$$\chi f'' + 2f' + f = 0 \qquad (11.27)$$

whose solution is $f = \chi^{-1/2} Z_1(2\chi^{1/2})$ where Z_1 is a linear combination of Bessel functions of order unity. The solution must be bounded as $x \to \infty$, and it follows that $\chi^{1/2}$ must be in the sense of $+i$, and that the desired solution is proportional to $(J_1 + iN_1)$. We can use known expansions to discuss some limits. For example we know that

$$\begin{aligned} J_1(s) + iN_1(s) &= 2i/\pi s + O(1) \quad \text{for } s \text{ small} \\ &= -(1 + i)(\pi s)^{-1/2} \exp is \quad \text{for } |s| \text{ large} \end{aligned} \qquad (11.28)$$

Putting $s = 2\chi^{1/2}$ we find, from the first, that the multiplier is $\pi\beta A_0$ so the final solution is $\psi = 2\pi A_0 \beta t \,(J_1 + iN_1)/s$, and the asymptotic solution is

$$\psi = -\pi^{1/2} A_0 \beta t \,\chi^{-3/4} \exp i\left(2\chi^{1/2} + \frac{\pi}{4}\right) \qquad (11.29)$$

This asymptotic solution is rather better than those of equations 11.23 and 11.24 for in the limit $|a\beta t| \gg 1$, it is valid for all values of x; a criterion concerning the time taken for a wave group of velocity $a^2\beta$ to cross the region of half-width a. Apart from this, the analytic solution is only partially satisfying, in view of the difficulty of calculating values of Bessel functions of complex argument, so it may be easier in practice to find f direct from equation 11.27 by numerical integration rather than to make use of the association with Bessel functions.

What we have here is I believe a hierarchy of approximations. While the similarity solution I find vaguely interesting, the Bessel function development seems academic but may have repercussions hidden from me. The moral that the labour involved in computing the value of an analytic expression may be greater than solving a problem numerically from the start is I believe fairly general.

11.11 Evolution of wavelike packets

The analyses so far start off with a spatially concentrated initial distribution, and follow it to a wavelike dispersion. What happens if we emphasise the other character, and start off with a fairly wavelike initial state? Consider the function

$$\psi = \int_{-\infty}^{+\infty} dk \, \exp{-a^2(k - k_0)^2} \exp{i(kx + \beta t/k)} \qquad (11.30)$$

At $t = 0$, $\psi = \exp(-x^2/4a^2) \exp{i k_0 x}$, which looks like a packet of sinusoidal waves so long as ak_0 is fairly large. Allowing for the term in ak to dominate the integral we can rewrite the exponentiand (!) as

$$-(a(k - k_0) - ix/2a)^2 - x^2/4a^2 + ik_0 x + i\beta t/k \qquad (11.31)$$

which suggests changing the independent variable to $\theta = a(k - k_0) - ix/2a$ and reduces the integral to

$$\frac{1}{a}\exp\left(-\frac{x^2}{4a^2} + ik_0 x\right) \int_{-\infty}^{+\infty} d\theta \, e^{-\theta^2} \exp\left(\frac{ia\beta t}{ak_0 + ix/2a + \theta}\right) \qquad (11.32)$$

which can be expanded when $|ak_0| \gg |x/a| \gg 1$ to give

$$\exp\left(-\frac{(x - \beta t/k_0)^2}{4a^2} + i(k_0 x + \beta t/k_0)\right) \qquad (11.33)$$

which is an envelope, travelling at the group velocity β/k_0^2, of waves with phase velocity $-\beta/k_0^2$.

Using the techniques of the previous section, we find that a long time later, if x and t are large, there is a large contribution to the integral only near $k^2 = \beta t/x$, which gives

$$\psi = \left(\frac{\beta t}{x^3}\right)^{1/4} \exp{-a^2\left(\left(\frac{\beta t}{x}\right)^{1/2} - k_0\right)^2} \exp{2i(x\beta t)^{1/2}} \qquad (11.34)$$

Again we see that a wave, of wavenumber k_0, whose phase velocity is negative has propagated to the positive value of x given by the group velocity.

11.12 Dispersion and dissipation

Dispersion of energy away from the region where it was released may be an important part of the mechanism of dissipation, for if the energy goes somewhere else it cannot be used to reverse an original localised process of energy release. Gravity waves confined to a box are likely to persist longer than when they are allowed to leak out into the surroundings. It is difficult to visualise dispersion in a Lagrangian sense. The active phase of a baroclinic wave can be interpreted as a systematic displacement of parcels resulting in a systematic decrease in their total potential energy. Parcels in the slantwise convection of Section 9.5 also have a significantly exponential character to their displacement, that makes their motion essentially non-reversible. What then are we to make of the dispersive phase of the motion that we think follows it? Do particle displacements in a dispersive wave have some essentially irreversible character?

The equation we have used here

$$\left(\frac{\partial}{\partial t} + U \frac{\partial}{\partial x}\right) \frac{\partial \psi}{\partial x} + \beta \psi = 0 \tag{11.35}$$

is perhaps the next simplest, after the advection equation, for testing numerical methods for the treatment of propagating waves, including as it does, advection but also simple but significant dispersion. Consistently, simple numerical solutions of this equation suggest that an allowable boundary condition is that 'nothing' propagates backwards, which we find astonishing, when all the waves do.

What is rather more disturbing is the possibility that, in a more general system, the fluid flow advects some properties into the area, while the group velocity propagates others out!

Chapter 12

Modelling

You pays your money, and you takes your pick.

12.1 Philosophy

An important aspect of scientific study is crystallised by the idea of a model. Having thought about a problem and gathered together all sorts of useful data, we begin to think we can see what is going on. In general the picture will be horrifyingly complicated, with many physical processes involved. For example, no model of large-scale atmospheric motion is likely ever to be able to reproduce the behaviour of individual cumulus clouds, but every model must include the vertical transport of heat, and possibly water vapour, by motion on that scale if it is to model properly the energy input to the free atmosphere.

12.2 Simulation

It may be that what we really want to do is to imitate the real world. This would be so if we wanted to sell our prediction to a user for example. We might then try to describe as many of the relevant processes as we could, usually in the form of equations, and using the best estimates of their parameters. This set could then be solved for appropriate initial conditions. Such a task would require the application of numerical methods and a large computer; hence the usual name of numerical model. When this simulation is executed using observed initial data, deficiencies become apparent. Some of these will be due to inaccurate

159

values for the parameters. For instance there was a great improvement in the weather forecast over England when the roughness of the land used in the model, was increased by a factor of 10 over what was previously thought reasonable.

Some error will be due to the vagaries of numerical methods. Step-by-step methods suffer from a tendency to explode, as in linear computational instability mentioned in Section 12.9. To counter this type of instability pseudo-viscosity is sometimes introduced. Having done this, one is not sure if this viscosity is supposed to be the effect of a real physical process, or part of the numerical integration scheme. A more spectacular instability is non-linear computational instability, mentioned in Section 12.10. Here there is feedback between the field being forecast and the velocity field doing the advection. This is so violent that unrealistically large pseudo-viscosity would be needed. When the viscosity is artificial we have no constraint on the appropriate law. It is often stated that ∇^2 is not selective enough and ∇^4 or greater power is better. A model allowed to take on more extreme values will often migrate towards them. Climate models usually head towards a state which no climatologist would recognise as being possible.

12.3 Academic modelling

It may be that we seek enlightenment, rather than imitation. We might then look for the essence of the physical system, disregarding all processes that we think are not absolutely essential. This set of essential processes defines our model, though perhaps we should call it a toy, rather than a model. The object is to reproduce the consequences of what we think are the essentials, and the only thing that is to prevent us is our inaccurate specification of the physics. What we can hope to get right is the feedback between processes. In the simulation model, this can get lost in the process of adjusting parameters to make the answer better, known as fine tuning. In the enlightenment model one would aim to get the physics right in extreme cases, perhaps at the expense of less accuracy in the middle of the range. Thus while present-day values may be more accurate, extreme values may be more reliable. A simple description of the radiative transmissivity of the troposphere at terrestrial wavelengths, which works well for zero moisture content and at saturation, may not work at all well for the middle range, and vice versa; extrapolation of an empirical formula which works well at present-day ranges may give a very wrong answer in extremes. For example a model gave small negative humidities in a dry region of the atmosphere. An answer within acceptable limits for some purposes, but we wondered what the transmission through a medium with a negative absorption coefficient might lead to.

For several reasons like this, the less simulatory, more 'first principles' models tend to be better at climate and worse at reproducing the present day. *sic* **model** ... proposed structure ... exemplary, ideally perfect ... **simulate** ... imitate, counterfeit ... deceptive ...

12.4 Mean planetary temperature

One very simple meteorological model concerns the estimation of the planetary temperature. Suppose that a fraction a (for albedo) of the solar irradiance S is scattered back into space. The remainder is absorbed by the planet and re-radiated to space. Assuming that the planet behaves like a perfect radiator, and noting that for a planet of radius r a disc of area πr^2 is exposed to the sun while the area re-radiating is $4\pi r^2$, thermal equilibrium demands that the planet attain radiative temperature T where

$$(1 - a)S = 4\sigma T^4 \tag{12.1}$$

where $\sigma = 5.67 \times 10^{-8}\,\mathrm{W\,m^{-2}\,K^{-4}}$ is Stefan's constant, and the values $a = 0.35$, $S = 1375\,\mathrm{W\,m^{-2}}$ give $T = 250\,\mathrm{K}$ which is a reasonable value for the re-radiating level; about halfway through the troposphere near the height at which water vapour becomes transparent in the terrestrial waveband. Several interesting speculations follow about the temperature of the surface and the state of water, principally that it is likely to change phase between vapour and liquid, with significant contribution to the heat budget through the latent heat.

12.5 Ice–albedo feedback

An interesting extension examines the feedback between ice cover and the albedo. Casual observation shows that the earth is more reflective over ice-covered regions than over bare earth. At some cool temperature we expect the planet to be just completely covered with ice, and therefore to have an albedo of about 0.62 which is typical of the regions of the planet currently icy. We estimate this critical temperature to be some $20\,\mathrm{K}$ cooler than present; this being the value that the present mean surface temperature exceeds that of the current ice-edge. At about $10\,\mathrm{K}$ warmer, the ice cap would just vanish to give the planet a bare-earth albedo of some 0.32. Suppose that the albedo varies linearly between these extremes, then

$$a = 0.62 - (T - 230\,\mathrm{K})/100\,\mathrm{K} \tag{12.2}$$

We can now solve to find the global mean temperature as a function of solar irradiance taking into account the ice-albedo feedback. This would help us to say what would happen to the planetary mean temperature if the input of solar energy were to change due to variation in the Earth's orbit, or volcanic dust or whatever.

Figure 12.1 shows the result as the bold curve. We notice several interesting features. It does not reproduce the result we had earlier for $a = 0.35$. A small adjustment to the albedo would correct this, but notice that this implies that the answer is rather sensitive to the value of the albedo. Indeed a change of 0.01 in a changes T by 1 K. When we look at the large spatial and temporal variations of albedo as seen from satellite pictures we begin to wonder if the atmosphere actually knows its albedo that well anyway.

We are astonished to see that over most of the range the temperature of the planet decreases as the solar irradiance increases. This can be traced to the insensitivity of the energy balance to the variation of radiation temperature;

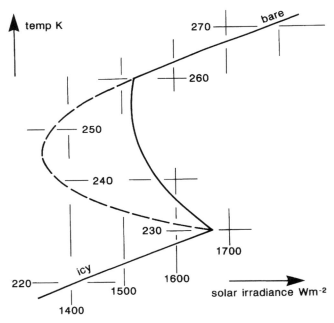

Figure 12.1 Variation of mean planetary temperature with solar irradiance. A simple zero-dimensional climate model relates mean planetary temperature T to energy balance and a plausible ice–albedo feedback. The thick line shows the consequences of assuming a linear variation of albedo with temperature. According to this increasing the irradiance causes a decrease in temperature. Such unreasonable behaviour needs to be explained. It is not good just to parameterise it away. Taking better account of the sphericity of the Earth gives the more reasonable picture represented by the dashed line. Nevertheless we are perilously close to the region of unreasonable and unstable climate.

σT^4 depends on T much less than does a. Thus if S is increased, the planet can achieve equilibrium only by increasing the albedo, which it can do only by becoming cooler. This does not sound like a proper thing for our model to do, and we can easily show that any albedo–temperature relation that gives the equilibrium temperature varying in the opposite sense to the solar irradiance will be unstable. Thus any state on the peculiar part of the curve will spontaneously jump to either the completely bare or the completely icy branches. If the planet became icy, S would have to exceed $1700 \, \mathrm{W\,m^{-2}}$ before the planet could get back to the ice-free state.

Improving the calculation of terrestrial radiation makes the problem worse, because the principal impediment to the escape of terrestrial radiation is water vapour, which we know generally increases with temperature like the saturation vapour pressure, so as to decrease the T^4 temperature dependence of outgoing radiation. Qualitatively, we argue that a warmer atmosphere contains more water vapour, so the radiation has to leave the atmosphere higher up, where it is cooler, so there is a degree of cancellation of the T^4 relation.

One might argue that the polar ice occupies a spherical fraction of the surface and that a better $a(T)$ relation would therefore be

$$a = 0.32 + 0.3((T - 260 \, \mathrm{K})/30 \, \mathrm{K})^2 \tag{12.3}$$

As shown by the dashed line in Figure 12.1, this gives the correct value for present-day albedo, and stable climates for $S > 1350 \, \mathrm{W\,m^{-2}}$.

12.6 Two-dimensional ice–albedo feedback

We like this model because we did not expect all the features it displays, so we are learning something. We now want to refine it, maybe by improving the $a(T)$ relation which it shows us is important. One way would be by using a 2-D model to calculate the latitudinal variation of temperature, specify a characteristic temperature for the edge of the ice sheet, hence the extent of the ice. We could use a plane geometrical model, which simplifies the mathematics , but in view of the probable significance of the geometry it would be much better to use spherical geometry. A simple but fairly convincing model assumes a constant eddy conductivity K to transfer heat between latitudes. The resulting equation is

$$K \frac{\mathrm{d}}{\mathrm{d}\mu} \left(\frac{(1 - \mu^2)}{R^2} \frac{\mathrm{d}T}{\mathrm{d}\mu} \right) = A + BT - \frac{1}{4}(1 - a) S Z(\mu) \tag{12.4}$$

where μ is sin latitude, $S Z(\mu)$ is the annual average solar radiation at the upper limit of the atmosphere, and the net terrestrial flux has been replaced by the linear approximation $A + BT$ where T is the deviation of the temperature from some suitable standard value. Notice that the albedo a changes from one value to the other at a critical temperature, T^* say. This dependence of a on T makes the equation non-linear in T, and is at the root of its complication and fascination. I would have guessed that such a low order of non-linearity would have trivial repercussions but in fact the results are quite complex. The parameters involved in this problem are:

$$a_0, \ a_1, \ R, \ K, \ B, \ T^*, \ Z(\mu) \tag{12.5}$$

This implies that the solution exists in something more than 8-D space: even a major visual, as well as computational task, to survey. However, some parameters represent only stretching of the scales and others an adjustment of the origin of scales. Taking this into account allows us to reduce the number of parameters.

First, we note that T^*, the temperature of the ice edge, is the only absolute temperature in the system, so it is natural to use it as the origin for the temperature scale and write $T = T^* + T_1 \theta$ where T^* is the base value, T_1 represents a scale for the temperature, and θ is now our non-dimensional temperature. Equation 12.4 becomes

$$\frac{\mathrm{d}}{\mathrm{d}\mu}\left((1 - \mu^2)\frac{\mathrm{d}\theta}{\mathrm{d}\mu}\right) = \frac{R^2(A + BT^*)}{KT_1} + \frac{R^2 B}{K}\theta - \frac{(1 - a)SR}{4KT_1}Z(\mu) \tag{12.6}$$

We can choose the temperature scale to be $T_1 = R^2(A + BT^*)/K$ which then accounts for the first term on the RHS by making its coefficient unity and leaving the parameters $\lambda^2 = R^2 B/K$ and $s = S/4(A + BT^*)$. The equation becomes

$$\frac{\mathrm{d}}{\mathrm{d}\mu}\left((1 - \mu^2)\frac{\mathrm{d}\theta}{\mathrm{d}\mu}\right) = 1 + \lambda\theta - (1 - a)s\,Z(\mu) \tag{12.7}$$

The remaining parameters are λ, a_0, a_1, s and $Z(\mu)$. The reduction from 8- to 4-dimensional parameter space is not negligible. The range of uncertainty in parameter values is roughly 0.02 in a due to error of observation and unrepresentivity, 0.2 in λ due largely to uncertainty as to the appropriate value to take for K and uncertainty in the form of $Z(\mu)$ of similar size, due to uncertainty in the appropriate average (equinoctal, annual mean, weighted by sunshine, etc.) that the term represents. These uncertainties allow a great range of possible solutions, including one close to that of Figure 12.1, but many of which are rather far away from the present climate. It is not clear to me what one is doing to the model if one demands that the parameters should take only those values such that the solutions do include the present. Nor is it clear from the results that doing so will substantially reduce the scatter. Boundary conditions are that there should be no heat flux at the pole or $(1 - \mu^2)\,\mathrm{d}\theta/\mathrm{d}\mu = 0$ there, which gives the problem an eigenvalue character. Incidentally we notice that integration of equation 12.7 with respect to μ from $\mu = -1$ to $\mu = +1$ gives an equation for the total energy balance, and reproduces equation 12.1 for the zero-dimensional model, with the average albedo used there replaced by the Z-weighted mean of a averaged between $\mu = \pm 1$. A lot of work for a rather small return.

12.7 Strategy in the use of numerical models

Tell me where is fancy bred, or in the heart or in the head?

How much more complicated it will be to diagnose a big General Circulation Model. It will be expensive to run, its progenitors may not always be anxious that the deficiencies of their model be exposed, and what if, at the end, there is an overriding integral constraint that tells us all we want to know, like that expressing conservation of energy for the whole system that we discovered in the 1-D climate model?

We might use our hierarchy of models to direct our line of research. For example what would we do if climate were controlled by the albedo through cloud instead of the ice–albedo? Almost certainly we would try to model cloud processes better. The most important cloud for determining the albedo may be the extensive sheets of layer cloud found at low levels near Indonesia. These may represent the accumulation of water vapour at the top of the boundary layer where large-scale subsidence warms the dry air above and so prevents deep convection followed by precipitation of cloud water leading to their dissipation. If this were true then there would be a dynamical-albedo feedback through the processes determining subsidence. Vertical motion of large scale is a very powerful generator of vorticity, so it may be that it is the vorticity budget, and particularly that generated by the large-scale eddies, that determines the albedo. We can visualise that such a model might be fundamentally different to the ones we have developed so far. A large model might not model the influence of subsidence on convection accurately enough.

12.8 Physics and numerical gridpoint models

Even for quite simple models, the way in which the arithmetic is done is important. The less we know about the form of the solutions, the more powerful and general needs to be their representation, and the more diverse and outrageous are the solutions the model can arrive at. Some aspects of fluid motion are represented by the simple wave equation

$$\partial^2 F / \partial t^2 + c^2 \, \partial^2 F / \partial x^2 = 0 \tag{12.8}$$

Much numerical analysis is tested out on the simple advection equation

$$\partial F / \partial t + U \, \partial F / \partial x = 0 \tag{12.9}$$

This is sometimes referred to in the literature as a wave equation but it is not a wave, but an advection equation. But what is so particular about the wave equation anyway? Certainly its solutions are no more wavelike than convenient solutions to many other equations – like the diffusion equation for example, and the wave equation will happily propagate many un-wavelike effects, like the shape of bricks, at least when there is no dispersion. If there is dispersion then things become much more interesting but then neither the simple advection equation nor the simple wave equation can represent them.

A powerful numerical technique replaces the continuous variation of the function by point values equally spaced in time and in space, together with a recipe for interpolation. Derivatives may then be expressed to any desired accuracy by finite-difference approximations. For example, if we use polynomials to interpolate between gridpoints we develop the finite difference approximation

$$\left(\frac{dF}{dx}\right) = \frac{F(x + \Delta x) - F(x - \Delta x)}{2\,\Delta x} + O(\Delta x)^2 \tag{12.10}$$

If we replace both the time and space derivatives this way, the advection equation becomes

$$F(t + \Delta t, x) - F(t - \Delta t, x) = (U\Delta t/\Delta x)(F(t, x + \Delta x) - F(t, x - \Delta x)) \tag{12.11}$$

This can be used, at least in principle, to forecast the function F forward in time, step by step. To see what would happen we can adopt a rather odd but powerful ploy and look for analytic solutions to the finite difference equation. By substitution we find that $F = A(k)\exp i(kx + \sigma t)$ is a solution of the difference equation 12.11 so long as

$$\sin(\sigma t) = (U\Delta t/\Delta x)\sin(k\Delta x) \tag{12.12}$$

This solution is quite general, because practically all initial distributions $F(x, t = 0)$ can be represented by a weighted sum of functions $\exp ikx$ so their temporal evolution can be too. A little manipulation shows that

$$\sigma = Uk\left(1 - \frac{(k\,\Delta x)^2}{6} + \frac{(k\,U\,\Delta t)^2}{6} + \dots\right) \tag{12.13}$$

so that the wave propagates at nearly the correct speed U, so long as $k\,\Delta x$ and $k\,U\,\Delta t$ are both less than about unity. This is already an interesting piece of information about redundancy of information, for it means that there must be about six points per wavelength and temporal period for reasonable representation of a sinusoidal function, which needs only the three parameters amplitude, phase, and wavelength for complete description.

We also note that the numerical solution is dispersive in the sense that the wavespeed varies with wavelength whereas the exact solution is not. We wonder if that dispersion is associated with the degradation of information that occurs during the numerical integration. There is one, but only one, other root to equation 12.12

$$\sigma = (\pi/\Delta x) - Uk + O(k\,\Delta x)^2 + O(k\,U\,\Delta t)^2 \qquad (12.14)$$

This wave propagates in the opposite direction to the fluid flow, and changes sign at each time step. This very unphysical nature brands it as a computational mode. We are astonished that there *is* a second solution, because the original differential equation was of the first order in time and so had only the progressive solution. What we have done is to replace a first-order differential equation with a second-order finite difference equation. We can tell because, for example, two initial conditions are needed to start off the time-stepping of the finite difference set. Another symptom is that the finite difference relates the 'last' to the 'next' value of the function, which also shows that it is of second order.

12.9 Linear computational instability

Sometimes, not only does the solution not move at 'the right' speed, but the amplitude also changes unphysically with time. Consider the waves which have $k\,\Delta x = \pi/2$, i.e. four points per wavelength. If $U\,\Delta t/\Delta x$ is greater than unity then the value of σ satisfying equation 12.12 must be complex and we see that the numerical method makes one of the two possible solutions amplify exponentially with time. That $U\,\Delta t/\Delta x < 1$ for stability is sometimes known as the CFL criterion, after the initials of the original discoverers. Since the shorter waves are most susceptible the phenomenon is also known as gridpoint instability. It must be avoided if physically acceptable results are to be obtained. This can be done by limiting the length of the timestep, but this becomes a severe limitation, particularly when sound waves are involved. Thus Δt must be less than 5 min for a 100 km grid. Fortunately these fast moving waves contain comparatively small amounts of energy so can be treated rather roughly. Sometimes massive dissipation, which acts mainly on the shortest waves, is used, or partly implicit integration schemes, or filtering through eliminating the physical processes that propagate the fast waves as described in Sections 4.8 and 4.13.

Much the same limitation is found for integration of the diffusion equation. Some of the ways we can stabilise the wave equation, like adding numerical diffusion, make the diffusion equation even more unstable, which raises problems if our system is governed by a balance between advection and diffusion.

We also note that, so far as pure advection is concerned, there is advantage in working close to the stability limit. Equation 12.12 is satisfied exactly for $U\,\Delta t = \Delta x$. Thus not only are the timesteps as large as possible, but the truncation errors are small. This is presumably because propagation is steady in the frame that moves one gridlength per timestep. As we will show in the next section this limit is a dangerous one.

We note one physical interpretation of the CFL criterion: that the timestep must not be longer than the time taken for information to arrive at the next gridpoint. This sounds like a nice visual physical interpretation, but if it were then it would surely be the group velocity, not the phase velocity that was relevant. I think that is not true.

In a diffusive system where information arrives instantly everywhere the information hypothesis would say that the numerical solution is unstable for all values of diffusion coefficient. Maybe it is. Interpolation always smooths the fields so numerical solutions to diffusion problems decay less quickly than analytic solutions. Now if we compare the analytic decay with the numerical decay we see that the numerical values increase exponentially *relative* to the analytic ones, so we might regard all such solutions as growing exponentially away from the analytic ones.

Maybe a better physical interpretation is much simpler and concerns only resolution of the temporally oscillating part of the solution. At four points to a period the central difference integration gives the sawtooth solution to the simply oscillation equation $d^2y/dt^2 = -\omega^2 y$ illustrated in Figure 12.2. For longer timesteps we see that the extrapolation gives an exponentially increasing amplitude. The real criterion is one of frequency of oscillation, not propagation of information.

$$\Delta t < \Delta x / U. \tag{12.15}$$

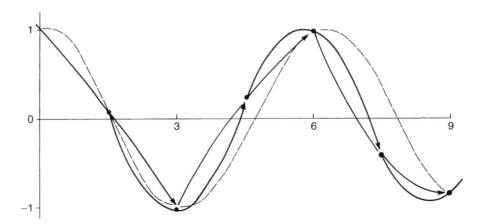

Figure 12.2 Linear instability. The centred finite difference approximation for $y'' = -y$ puts a parabola through the last two points and uses the curvature from the differential equation. The numerical solution is therefore a set of interlaced parabolas, not a continuous curve. As we get to only four points per wavelength the representation becomes very crude.

The simple time-centred difference scheme for the advection equation has two solutions, one of which essentially changes sign at each timestep and may be unstable. Can we modify the numerical scheme so that it recognises only the physical solution? One way is to write the finite difference approximation so that it connects only $F(t + \Delta t)$ and $F(t)$ which makes it a first-order difference, as well as differential equation. This scheme is correct only to the first order in Δt, unless we use the information at the new timestep to refine our estimate of the change during the timestep. This makes a nice, simple integration scheme; one of the Runge–Kutta family.

An alternative ploy predicts new values of the function at points halfway in space between the old ones. This 'staggered grid' scheme has a similar effect. Yet another way is to average the field every now and again to remove the shorter waves. This smoothing can be done in space or in time and is equivalent to putting additional computational smoothing in the system. It usually represents computational dissipation and may have a significant effect on apparent energy conservation. Finite difference schemes already represent numerical smoothing. Expanding the finite difference approximation to dF/dx in equation 12.11 gives

$$\frac{F(x + \Delta x) - F(x - \Delta x)}{2\Delta x} = \frac{dF}{dx} + \frac{(\Delta x)^2}{3!}\frac{d^3 F}{dx^3} + \cdots \qquad (12.16)$$

showing that we have done something like introduce diffusion of size $k^2 \Delta x^2/6$ into the model. All these schemes have their own individual criteria for stability, and truncation errors.

12.10 Non-linear computational instability

More spectacular is an instability that sometimes seems to happen in just one timestep. The answers look alright, if a little rough at one timestep, and have disappeared out of range of the computer by the next. This is noticed with non-linear equations, of which one of the simplest is this simple extension of equation 12.9

$$\frac{\partial u}{\partial t} + u\,\frac{\partial u}{\partial x} = 0 \qquad (12.17)$$

Replacing the derivatives by centred finite differences as in the linearised problem, we get a finite difference prediction equation with two parameters. Writing U as the space-average of u, these are $U\,\Delta t/\Delta x = a$ and e, the magnitude of the variation of u about U along the wave. Integrating forward in time we find that, for the CFL parameter a sufficiently small and e not too big, the scheme is stable. But that for values of a near to but less than unity and e rather large the scheme is unstable. The analytic equation 12.17 conserves u^2 averaged

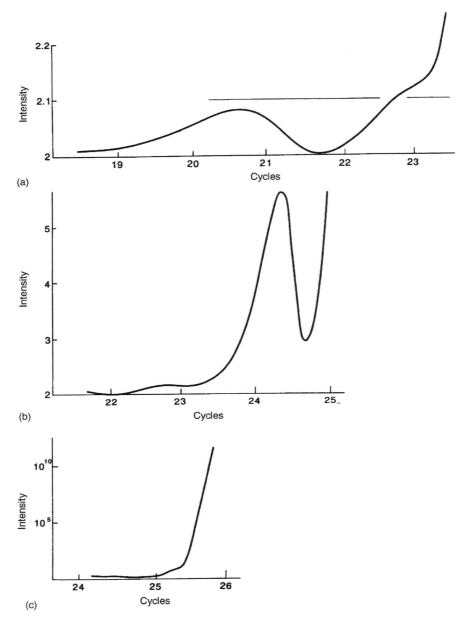

Figure 12.3 Non-linear instability. A variable that should be conserved according to the physics (x^2) is propagated by a non-linear wavelike equation working near the linear instability limit. Notice the change of scale between successive diagrams. (a) While the first 17 waves pass through the system, the numerically predicted value varies from the true value 2 by less than 0.005, but (b) at 21 periods there is a great surge to 2.07 which remarkably, dies down again, until at 24 periods the energy increases to 5.8, pauses, dies down to 2.9 for one period or so before (c) increasing to 10^{10} in the next cycle (with a logarithmic scale for the intensity).

in space. This quantity looks a bit like kinetic energy, and is a useful quantity to monitor. Figure 12.3 shows its evolution. We have chosen four points per wavelength as being perhaps the most susceptible to instability, and certainly the smallest number of points we can reasonably expect to have to integrate. Three points would have sufficed, but with their less virulent linear instability, did not seem quite the best. We chose to work rather near the linear stability limit with $U \Delta t / \Delta x = 0.99$ and a superposed wave of amplitude $a = 0.1 U$. The linearised equation is stable, and the integration is rather accurate for time integration much larger than we show for the non-linear version, because the criterion is so close to the stability limit as mentioned above. This checks that numerical truncation error is not important. We then ran the non-linear version with results as illustrated in Figure 12.3. While the first 17 waves passed through the system, the numerically predicted value for the energy varied from the true value 2 by less than 0.005. At period 21 there is a great surge to 2.07 which remarkably dies down again, until at period 24 the energy explodes to 5.8, pauses, dies down to 2.9 for one period, then goes to 10^{10} in the next cycle, shown in the last window using a logarithmic scale. Thus, the system behaves nicely for some considerable time, but takes more and more vioent excursions about the correct value before exploding spectacularly. In practice we have to imagine that this solution is initially submerged in a large meaningful component of longer wavelength, in which it represents 'noise' and from which it emerges as the pronounced instability develops.

From Figure 12.3(c) we see that the growth is much more rapid than exponential. We suspect that it may be multiplicative in character, with the amplification being doubled by the non-linear terms at each time step. Study of this last picture shows this is not far from the case, with the multiplication at each timestep perhaps being nearer 1.7 than 2. We wonder, when did the instability start? The solution at 22 periods 'looks' much the same as it did at any time up until then. Why did the scheme wait so long and what finally set it off?

Chapter 13

Models

13.1 Types of models

Analytic methods work best when the important variations are localised, and so are easier to isolate often in the form of mathematical singularities. Numerical methods work best when the variation is smooth and continuous, as in polynomial functions of low order, and therefore easier to resolve. Combinations of analytic and numerical models are sometimes appropriate. First we explore some models of large-scale flow in which the vertical variation is expressed using only two parameters but which have great resolution in the horizontal. We go on to look at some of the consequences of severe truncation in the horizontal representation.

Suppose that we are satisfied that the incompressible, quasi-geostrophic vorticity equation will display some interesting features of flow between two rigid lids. Our defining equations are, as in Chapter 9,

$$\frac{DZ}{Dt_h} + \beta v = f \frac{\partial w}{\partial z} \qquad \frac{D\phi}{Dt_h} + Bw = Q$$

where

$$\mathbf{v} = \mathbf{k} \wedge \nabla \psi, \quad Z = \nabla^2 \psi, \quad \phi = (f/g)(\partial \psi/\partial z) \tag{13.1}$$

13.2 Models with two levels

There are many different ways of treating the vertical variation in this equation. We could use the values of ψ on the two surfaces $z = 0$ and $z = H$. To satisfy the boundary condition $w = 0$ on these surfaces, standard numerical technique suggests we also carry the next values outside this range $z = -H$ and $z = +2H$ which gives a nice centred difference approximation to $\partial \psi / \partial z$ at each boundary.

Alternatively, we might specify ψ at $z = H/4$ and $z = 3H/4$, and w at $z = H/2$; a vertically staggered grid. The ψ field then satisfies

$$\left(\frac{D}{Dt}\right)_1 \nabla^2 \psi_1 + \beta \frac{\partial \psi_1}{\partial x} = +f \frac{2w_{1/2}}{H}$$
$$\left(\frac{D}{Dt}\right)_3 \nabla^2 \psi_3 + \beta \frac{\partial \psi_3}{\partial x} = -f \frac{2w_{1/2}}{H} \qquad (13.2)$$

at $z = H/4$ and $z = 3H/4$ respectively. Finally, conservation of potential temperature at the middle level $z = H/2$, gives a third equation connecting ψ_1, ψ_3 and $w_{1/2}$

$$\frac{f}{g}\left(\frac{D}{Dt}\right)_2 \left(\frac{\psi_3 - \psi_1}{H/2}\right) + B w_{1/2} = 0 \qquad (13.3)$$

so long as we use use the mean stream function $\psi_2 = (\psi_1 + \psi_3)/2$ in the substantial derivative $(D/Dt)_2$. How can we judge how good this equation set is? Perhaps we can select a test case which is physically important and for which we know the correct answer, and compare solutions with that. It is of little use to select a test case merely because we can write down an analytic mathematical solution for it, for such solutions are likely to have special mathematical properties that make them unrepresentative of important physics. I am therefore suspicious of tests using solutions defined by $\nabla^2 \psi = F(\psi)$ or similar formulations, which appear to test mathematical niceties rather than physical processes.

Suppose we choose to examine the accuracy of the equations in describing small amplitude, adiabatic perturbations, of uniformly sheared zonal flow, to see how well they represent the classical problems of amplifying baroclinic and neutral dispersive barotropic waves. Putting

$$\psi = -Ay(z - H/2) + \epsilon \exp ik(x - ct) \qquad (13.4)$$

where ϵ is suitably small, we get

$$ik^3(c + AH/2)\epsilon_1 + ik\beta \epsilon_1 = +2fw_{1/2}/H$$
$$ik^3(c - AH/2)\epsilon_3 + ik\beta \epsilon_3 = -2fw_{1/2}/H \qquad (13.5)$$

and

$$2ikc\frac{(\epsilon_3 - \epsilon_1)}{H} + \frac{A}{2}ik(\epsilon_1 + \epsilon_3) - \frac{N^2}{f} w_{1/2} = 0 \qquad (13.6)$$

The algebraic manipulations are tedious, but can be ameliorated a little by paying attention to symmetries and to pleasing groups of symbols that become evident. Mine passed through the stage

$$\left(k^2c + \beta + \frac{4f^2c}{N^2H}\right)^2 - \left(\frac{AHk^2}{4} - \frac{Af^2}{N^2H}\right)^2 + \left(\frac{Af^2}{N^2H}\right)^2 - \left(\frac{4f^2c}{N^2H}\right)^2 = 0$$

(13.7)

where A represents the baroclinity, and the combination $f^2/N^2H^2 = k_*^2$ is evident. We can write the solution in the form

$$(k^2c + \beta)(k^2c + \beta + 8k_*^2c) + k^2(k_*^2 - k^2)A^2H^2/16 = 0$$

(13.8)

This displays several interesting features, though it must be said that we might not have chosen to write it this way if we did not know what we were trying to see. Baroclinity is represented in equation 13.8 through the thermal wind and therefore the shear, by the term in A. Barotropic conditions are given by putting $A = 0$. We then find that the two roots of equation 13.8 represent the analytic solutions for Rossby waves with zero and one node in the vertical and we cannot expect more solutions than that from a two-level model.

The baroclinic term always vanishes for $k = k_*$ and shorter waves are unstable, so this model predicts a fixed short-wave cutoff for instability. Analytic solutions valid for short wavelengths, as in Section 9.9, show that there is not an exact short-wave cutoff to growth but that there *is* a wavelength where the growth rate is small compared to the maximum, and that this does not vary much. The phase speed of very short waves is $\pm AH/4$ where the analytic solution gives $\pm AH/2$. We see in Figure 13.1 that the asymptotic steering level is at the extreme grid level rather than at the boundary, which seems not unreasonable. The baroclinic aspects of the solution shows up most clearly for $\beta = 0$, shown in Figure 13.2. It represents the growing waves rather well, with for example maximum growth rate only 5% too small.

The mixed solution, when barotropic and baroclinic effects compete, is nessesarily more complicated, but the solution for the typical value $\beta = AHk_*^2$ shows reasonable agreement. We should ask 'reasonable compared with what' and perhaps compare this solution with that using another model. It may be that all models will give the same answer because it is too easy a question we are asking. We therefore show one solution using the finite difference scheme in which the flow is expressed in terms of stream functions at $z = 0$ and $z = H$ with dummy values at $z = -H$ and $z = 3H$. The complex phase velocity is given by

$$k^2(k^2 + 4k_*^2)\left(c^2 - \frac{A^2H^2}{4}\right) + 2\beta(k^2 + 2k_*^2)c$$

$$+ \beta^2 + 2A^2m^2k^2 - \frac{A^2H^2}{4}k^4 = 0$$

(13.9)

This, with the same value of β as above gives the fastest growth rate less well, though the short waves now have the correct steering level.

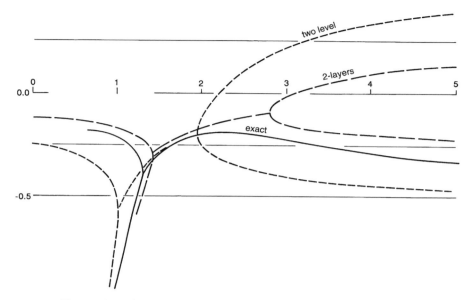

Figure 13.1 Phase speed of baroclinic waves, with the realistic value
$\beta = AM^2/h$. *2-layer; large dashes, 2-level; shortdashed, exact; continuous*

It is important to remember that our judgement is subjective. We have supposed that the rate of growth of the waves might be a good test. But one could well argue that since the eddy spends only 20% of its lifespan amplifying it does not really matter if we get the duration of that part of its lifetime a little wrong. Perhaps it is the processes that determine the final amplitude that are critical, in which case we might look at the energy conservation properties of the scheme. It might be necessary to consider the processes of dissipation more carefully. Surface friction is likely to be important in dissipation, but surface conditions are poorly represented in a model with horizontal velocities represented only at levels $z = H/4$ and $z = 3H/4$. Perhaps levels $z = 0$ and $z = H$ would be better for this purpose. It might be important to represent well the dispersive phase of the motion associated with momentum transfer and the maintenance of anomalies of large scale. This is likely to be at a stage in the life cycle when the surface energy has been reduced by friction and so the dispersive character of the upper-level flow is what needs to be modelled well, in which case we might again prefer levels $z = 0$ and $z = H$.

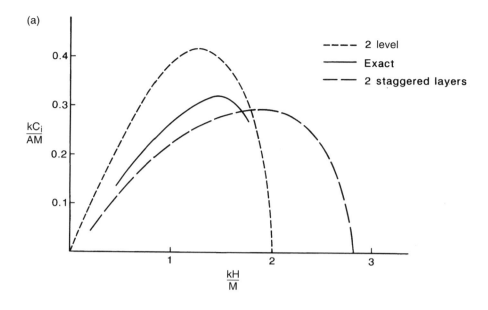

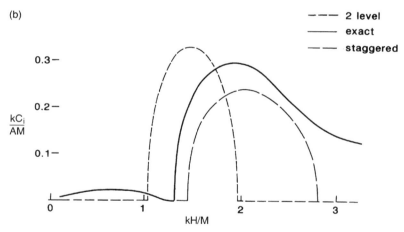

Figure 13.2 Growth rate of baroclinic waves with (a)$\beta = 0$ and (b)$\beta = AM^2/H$, according to several 2-parameter vertical representations. Solutions are qualitatively similar, but with substantial differences in wavelength of maximum growth and maximum growthrate.

13.3 Two layers

Suppose that the continuous variation of potential density with height is replaced by two homogeneous immiscible layers with the height of the interface between the two layers allowed to vary with horizontal position and time. The equations are similar to those for the two-level model, at least if the height of the interface varies little over the domain, and both schemes are in trouble if it does not.

Oceanographers find the two-layer model attractive because it does look rather like the thermocline dividing up the real ocean. But one of the most interesting parts of the real ocean is where the thermocline intersects the surface. This is because the transfers of heat, gases, and momentum, to the atmosphere, and the baroclinic transfers from the mixed layer into the deep ocean are large there. Thus carbon dioxide, and phosphates, and warmth get mingled in this zone to generate transfers and biological processes of importance to global scale balance. Unfortunately this is just where the two-layer model, as usually formulated, is least reliable.

An advantage of the two-layer scheme is that all the variables are defined everywhere, even though the discontinuous interface region is physically unrealistic because it is technically unstable to shearing instability, and the discontinuity is not consistent with diffusive transfer.

Basic equations for the two-layer quasi-geostrophic approximation are easily derived. We suppose two layers of constant potential density in each, separated by an interface at height h. The term $f \, \partial w / \partial z$ in the vorticity equation is important. The vertical velocity at the top of the lower layer is $(Dh/Dt)_0$, where the subscript zero is to indicate that advection is by the velocity of the fluid in the lower layer. With no horizontal variation of potential density within the layer the vertical velocity must vary linearly with height to vanish at the lower lid, and the vorticity equation 13.1 becomes

$$\left(\frac{D}{Dt}\right)_0 (\nabla^2 \psi_0 + \beta y - \frac{f}{h} \, \delta h) = 0$$

$$\left(\frac{D}{Dt}\right)_1 (\nabla^2 \psi_1 + \beta y - \frac{f}{H-h} \, \delta h) = 0 \qquad (13.10)$$

for the lower and upper layer, respectively and where $\psi_1 - \psi_0 = g' \, \delta h / f$ and $g' = g(\rho_0 - \rho_1)/\rho$. This set is similar to the two-level equations, with appropriate reinterpretation of the parameters. For example, solving the linearised stability equations, we find that the critical wavenumber for stability is hf^2/g' but the best value of g' to take to represent a state of continuous stratification is open to debate and again depends on the intended application. We notice that the quantity $\nabla^2 \psi + \beta y - f \, \delta h / h$ represents the relative vorticity corrected for the effect of movement relative to the underlying Earth and vortex stretching; the quasi-geostrophic potential vorticity for a two-layer model. It is salutory that the vertical velocity is not in general continuous at $z = h$, because the horizontal component of velocity is different on each side of the interface.

13.4 Two parameters

In this system, the vertical variation of the stream function is represented by a continuous function of the vertical coordinate with two coefficients. It is consistent to use a quadratic function for the vertical velocity between two rigid lids. We represent the stream function, and the vertical velocity by

$$\psi = \psi_0 + z\,\psi_1 \quad \text{so} \quad \phi = \frac{f}{g}\,\psi_1 \quad \text{and} \quad w = 6\overline{w}\left(\frac{z^2}{H^2} - \frac{1}{4}\right) \tag{13.11}$$

for $-H/2 < z < H/2$, where ψ_0, ψ_1, and $\overline{w}$ are functions of the horizontal coordinates and time, and the functions 1 and z are orthogonal to each other over the range $-H/2 < z < +H/2$. We now demand that the vertical average which is the zeroth moment, of the vorticity and potential temperature equations are satisfied

$$\left(\frac{\partial}{\partial t} + \mathbf{v}_0.\nabla\right) Z_0 + \frac{H^2}{12}\mathbf{v}_1.\nabla Z_1 = 0$$

$$\left(\frac{\partial}{\partial t} + \mathbf{v}_0.\nabla\right) \frac{f}{g}\,\psi_1 + B\overline{w} = 0$$

where

$$Z_0 = \nabla^2\psi_0 \qquad Z_1 = \nabla^2\psi_1 \tag{13.12}$$

A second relation is given by the first moment with respect to height of the vorticity and potential temperature equations

$$\left(\frac{\partial}{\partial t} + \mathbf{v}_0.\nabla\right) Z_1 + \mathbf{v}_1.\nabla Z_0 = -\frac{12f}{H^2}\overline{w}$$

$$= \frac{12f}{HN^2}\left(\frac{\partial}{\partial t} + \mathbf{v}_0.\nabla\right)\psi_1 \tag{13.13}$$

This scheme gives accurate answers for the instability problem, but does not have an analogue for potential vorticity. Some choice remains in the choice of functions for the vertical distribution. While polynomial functions of height seem plausible for the incompressible model, there is scope for change when the variation of density is included, and similar considerations arise when stratospheric processes are important, and comparatively great height ranges are needed.

13.5 Layer aspect of models

We have chosen rather crude models, with only a very few layers in the vertical in order to explore their limitations. All are capable of simple extension, but what have we learned from the simplest cases? First we must decide on what criterion we will use for our judgement. While all models seem to represent baroclinic instability with about the same fidelity, the two-parameter model does rather better, but does not have an analogue to potential vorticity. If this property is important to our intended application, then the value of this formulation is reduced.

Correct solutions of the stability problem show one curious, feebly amplifying long wave which is not one of the barotropic solutions one expects at long wavelengths. There are also extra waves near the critical wavelength near where these new amplifying solutions start. These solutions do not occur in the models with few vertical parameters. It is generally believed that these waves are unimportant, but they do have phase velocity close to the velocity of the low-level air, so the waves are nearly stationary, and might resonate with the stationary topographic forcing.

Finally we notice that the numerical solutions represent best the vigorous waves, and associate that with their lack of 'criticality'. When the waves amplify slowly they behave violently close to the steering level. This is difficult to represent using widely spaced gridpoints. In contrast the mathematical analysis of which Section 9.9 provides one example, is most powerful when there are such singular points in the region, for they serve as origin for expansions. This leads to much literature on 'the just-neutral' solution which I think is of marginal interest to the hearty growth of a lusty baroclinic wave.

13.6 Spectral models

While the use of functional representation in the vertical is debatable, several distinct advantages come from the use of orthogonal functions for the horizontal variation. For geometries with no essential singularities there is no system as nice as the grid of rectangular Cartesians. However, where the coordinate system is singular we find over-representation and artificial singularities at $r = 0$ in the cylindrical system, at the polar axis in spherical coordinates, as in the simple climate model of Section 12.6. Use of orthogonal functions avoids this difficulty, requiring (almost) only that the weighting function defining orthogonality be specified.

When the whole system is being treated it is reasonable to suppose that area is a suitable weight. In a Cartesian domain this might well give sines and cosines of multiples of the two horizontal coordinates as a useful set. The weighting function for cylindrical coordinates would be the radial distance, and Bessel functions of radius together with sinusoidal functions of azimuth angle arise.

Use of orthogonal functions allows us to arrange a precise cut-off for kinetic energy. We can arrange the equations so that energy does not leak out of the system just because of misrepresentation of the non-linear terms in the advection. Because there is no great unphysical energy leak, crude spectral models seem to behave more interestingly than crude gridpoint models. But as the resolution of spectral models is made finer, by including more terms in the representation, the advantages over gridpoint models of a similar complexity seems to decrease. The volume of arithmetic increases more violently with increasing resolution than it does for gridpoints. Energy begins to accumulate at the short-wavelength end of the spectrum, and this has to be removed, for instance by introducing scale-selective diffusion to take it away. Then, apart from having overcome the not inconsiderable geometrical problems, we are back again with an energetically leaky system.

A disadvantage of the spectral over the gridpoint-advection schemes is that unattainable values of variables may appear. While this is usually acceptable as a measure of the accuracy of the scheme, we can visualise trouble. For example, terrestrial radiation is absorbed and emitted by water vapour, so in regions where the relative humidity was represented as negative there might be trouble. Similarly, cloud microphysics runs on tiny supersaturations, so, faced with a region with a relatively enormous supersaturation, there might be a catastrophy.

13.7 A simple spectral model

A simple but instructive spectral model shows some aspects of the evolution of baroclinic instability. Consider uniformly sheared flow in a channel of great length, and finite (scaled) width π, with $\beta = 0$. We know that the flow is unstable and that wavelike perturbations will amplify with time. To the first order in amplitude of the perturbation the solution can be written as

$$\psi = -yz + (A \sinh nz \, \cos \lambda x + B \cosh nz \, \sin \lambda x) \sin y$$

where

$$n^2 = 1 + \lambda^2 \qquad\qquad (13.14)$$

We find that this expression satisfies the vorticity equation of Section 9.6 exactly, but that the equations expressing the boundary condition at the lids generate extra terms in

$$\sin 2y, \; \sin 2\lambda x \sin 2y, \; \cos 2\lambda x \sin 2y, \; \sin \lambda x \sin 3y, \; \cos \lambda x \sin 3y$$

$$(13.15)$$

Our 'simplest' truncation scheme adds only the term $C \sin 2y \cosh 2z$ to ψ. This enables us to eliminate the error term in $\sin 2y$ in the non-linear boundary equation. This truncation in wavenumber space defines our 'simplest' non-linear spectral model.

The truncation may seem somewhat arbitrary at first sight, but the term in $\sin 2y$ represents the decrease in the mean baroclinity, which is what is changed by the transfer of thermal energy by the wave, and which we believe to be an important aspect of the energetics.

The terms in $\sin 3\lambda x \sin y$ and $\cos 3\lambda x \sin y$ show differences between the cyclonic and anticyclonic parts of the wave, which is something we are not concentrating on just here, comes only as a pair so needs two new coefficients, and does not represent a change in the energetics. As it is, the equations that define the vertical velocity are already quite complicated. We have

$$\frac{D}{Dt} = \frac{\partial}{\partial t}$$

$$+(z - A \sinh nz \cos \lambda x \cos y - B \cosh nz \sin \lambda x \cos y - 2C \sinh 2z \cos 2y)\frac{\partial}{\partial x}$$

$$+(-\lambda A \sinh nz \sin \lambda x \sin y + \lambda B \cosh nz \cos \lambda x \sin y)\frac{\partial}{\partial y}$$

and

$$\frac{\partial \psi}{\partial z} = (-y + nA \cosh nz \cos \lambda x \sin y$$

$$+ nB \sinh nz \sin \lambda x \sin y + 2C \cosh 2z \sin 2y) \qquad (13.16)$$

We require the spectrally truncated version of equation 9.23 to vanish at $z = \pm 12$. We notice that the condition at $z = +1/2$ implies that at $z = -1/2$, which is fortunate otherwise we would need twice as many coefficients to satisfy our simplest system. The coefficients of $\cos \lambda x \sin y$, $\sin \lambda x \sin y$, and $\sin 2y$ in equation 13.16 are

$$n\cosh\frac{n}{2} A' + (\frac{n\lambda}{2} \sinh\frac{n}{2} - \lambda \cosh\frac{n}{2}) B + (n\lambda \sinh\frac{n}{2}\sinh 1 - 2\lambda \cosh\frac{n}{2}\cosh 1) BC$$

$$n\sinh\frac{n}{2} B' + (\lambda \sinh\frac{n}{2} - \frac{n\lambda}{2}\cosh\frac{n}{2}) A + (2\lambda\cosh 1 \sinh\frac{n}{2} - n\lambda^2 \sinh 1 \cosh\frac{n}{2}) AC$$

and

$$2\cosh 1 \, C' + \frac{n\lambda}{2} AB \qquad (13.17)$$

where the prime indicates differentiation w.r.t. time. Requiring these coefficients each to vanish gives our set. While the quantities A, B, and C, are small the vanishing of coefficients 13.17 reproduces the linear waves whose amplitude varies exponentially with time. This is not surprising since that is the way they were constructed. For larger amplitudes the term in $\sin 2y$ becomes significant, showing that the mean potential energy has been changed. Multiplying the first coefficient by A the second by B and using the third, we notice that, with suitable constant multipliers, the three expessions add to zero showing that there is an integral constraint of the form

$$\alpha A^2 + \beta B^2 + \gamma C = constant \tag{13.18}$$

where *constant* is independent of time. We identify the first two terms as the total kinetic energy in the system, and therefore associate the third term with the potential energy that is available for conversion into kinetic. In fact, this was the first stage of the route by which I came across the concept of available potential energy. The generalisation of Section 5.1 came later. Margules knew about it all the time of course.

We would like to ask what the general solution of the set of equations given by 13.17 looks like. We could scale A, B, C, and time. This would reduce the original set to one with one parameter, though there may be options as to the direction of time where the square root of negative scaling factors would otherwise be involved. Thus it seems to be a little more than a four-parameter system; one from the scaled differential equation and three more from the starting values of A, B, C. However, many solutions will initially be part exponentially growing and part exponentially decreasing, so more-or-less dominated by the amplifying solution. When we are in the parameter range for which there is no amplifying solution the waves will presumably remain of small amplitude throughout. Recollecting the odd behaviour of non-linear instability, we may not feel so sure.

Solutions can be derived analytically, in terms of elliptic-integral functions. These behave like $1/\cosh \gamma(t - t_0)$ for some ranges of parameters. Thus we see the exponentially amplifying phase where $(t - t_0)$ is large and negative, followed by maximum amplitude at $t = t_0$, then exponential decay back to the initial state.

While we can solve these equations analytically, it is tedious, and largely numerical anyway, so there are advantages in solving numerically from the start.

13.8 Some non-linear solutions

The fastest-growing wave for this width of channel is for $\lambda = 1.5$ nearly. Figure 13.3 shows the evolution to finite amplitude of this solution. At first kinetic energy increases nearly exponentially with time, with a constant phase lag between the stream function at the top and bottom of the layer. As the amplitude approaches the maximum, the phase changes over to the opposite sense, and the kinetic energy decreases as the wave kinetic energy passes back into potential. Two integral constraints that should be satisfied by the exact three differential equations 13.17 are indeed closely satisfied, which gives us confidence in the numerics. Like a beam-pendulum, initially suspended with its

(a)

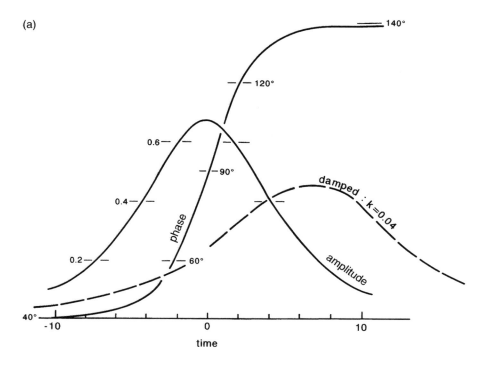

(b)

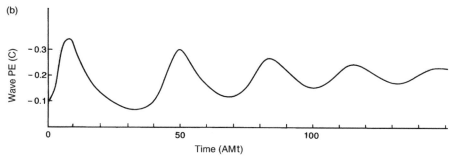

Figure 13.3 (a) Evolution to finite amplitude for the simplest non-linear Eady wave. Continuous lines show the amplitude increasing, at first exponentially, then the phase between top and bottom changes sign and the amplitude decreases to zero again. The dashed line shows that the amplitude increases more slowly with friction, but does not go to zero at large time. (b) The wave potential energy (C in equation 13.14) , now seen on a rather longer timescale does not decrease to zero, but oscillates about a non-zero value. We speculate on the nature of the air coming out of mysterious boundary layers.

centre of gravity above the pivot point, the system has accelerated exponentially away from the unstable equilibrium, acquired kinetic energy at the expense of potential, then converted it back again as it swung up the other side. If the system had a not-quite exponential solution, by being started off at small displacement but with zero initial velocity for example, it would by analogy with the pendulum be periodic, but very non-sinusoidal as it clambered back, like Pooh bear, up the other side. We can verify that our equation is periodic, but it is not very interesting meteorology, or is it just an example of incipient unpredictability? No doubt the model would be made more realistic if allowed more wavelengths for the energy to explore, but are there interesting experiments that can be done using the present easy truncation?

13.9 Non-linear baroclinic wave with friction

Friction at a boundary induces vertical velocities correlated with vorticity, as described in Section 8.4. Doing this, realistically and importantly, at only the lower boundary destroys the fortunate symmetry that allows us to use only half the terms in $\sinh z$ and $\cosh z$ that we expected. To preserve this simplification we must have the same friction at top and bottom. This artificiality reduces the value of the exercise. However the equations used for the calculation so far become

$$2.589A' - 0.762B - 3.370BC = -3.33kA$$
$$1.848B' - 0.394A + 0.204AC = -4.64kB$$
$$3.086C' + 1.347AB = -4.70kC \qquad (13.19)$$

where k is a non-dimensionalised version of the constant in the friction-layer boundary condition. We see that all three friction terms look properly dissipative in the sense of each on their own imply exponential decrease with time to A, B, and C. Integration using the realistic value $k = 0.04$ shows in Figure 13.3(a) that, though the amplification rate is substantially decreased, by about as much as we would expect from the linear equations, the wave essentially just takes longer to get to a modestly decreased amplitude. More surprising perhaps is that Figure 13.3(b) suggests that the potential energy of the wave, represented by C does not die out as rapidly as it did before, in spite of the friction we added. Indeed we can see the possibility of an ultimate steady eddy state actually supported by friction. Thus one of the energy-like relations including the friction is

$$\frac{\mathrm{d}}{\mathrm{d}t}(1.29\,A^2 + 15.3\,B^2 + 16.7\,C) = -k\,(3.33\,A^2 + 76.6\,B^2 + 25.4\,C) \qquad (13.20)$$

The original source of potential energy is the baroclinity, represented by the term in yz in ψ, and by the coefficient C. We now see from Figure 13.3(b) that the friction, by diminishing the coefficient C serves to inject energy into the system. Perhaps this might be because we have implied some unrealistic property to the air which appears out of our mysterious boundary layer?

13.10 What do we learn from such exercises?

Some of the effects we have discovered may appear in more complex models, perhaps heavily disguised. Some of the constraints, like the symmetrical boundary layer and the violent truncation, suggest that our model is over-constrained by the desire for mathematical elegance. Even more constrictive is the demand that all the physics of the fluid between the boundaries should be summarised by the simple Laplacian condition on the stream function. The fact that our simple lateral boundary condition makes the vertical velocity vanish everywhere along the side wall is another disquieting feature not necessarily confined to this simple model.

As it is, the present model does several things that are surprising. Because it is so simple one can begin to see where these curious effects have come from. Perhaps it is not easy. We can construct more complex models, like ones containing a much larger set of wavelengths, but perhaps these effects will still be there now hidden under much more detail. We find that the complexity of multiplying two Fourier series together, which we can glimpse from the very simple case with just three coefficients, becomes overbearing. But that it is more economical to calculate the results of the advections in Cartesian space then Fourier analyse the result to get it back into wavenumbers. We also find that the energy tends to get held up at the truncation-edge of the spectrum, and special numerical devices must be used to remove it.

Or we can ask, why are we asking questions; what it is that we want to know? Suppose we would like to predict climate. The precise details of the evolution of a particular system are of less interest than the way such systems interact with each other over their lifetimes. One way of tackling this problem is by parameterisation. We declare how we think a system should behave and then study the consequences of having this conceptual system interact with some other. This emphasis on the feedback between different scales produces a somewhat different philosophy.

Chapter 14

Transport and mixing

14.1 Transfer

We are concerned here with the transfer of properties from one place to another. This transfer may be by bodily contact as when momentum is transferred from one airstream to another by the action of pressure forces. Transfer may be by pure advection with the bulk motion of the fluid, as when water vapour is carried along at constant mixing ratio together with the dry air. Such simple advection may be complicated by the relative motion of a different phase as when water droplets fall relative to the air. Molecular diffusion may be important, as when warm air is brought close to cold air then brought back to where it started, but cooler. To form a cloud droplet, water vapour diffuses towards a cloud particle and the latent heat released is conducted away from a hot particle. Electromagnetic radiation transfers energy from one absorber (emitter) to another, and sound waves carry pressure information.

Here we concentrate on the advection by fluid motion, but find that we must be aware of these other processes at the same time. For example, when the fluid motion is inexactly known, the statistical fluctuations round the known state may resemble diffusion in some respects, but not in others. It is necessary at least to be aware of the possible differences. The transfer of a variety of properties by fluctuating unresolved, but *not* necessarily random, fluid motion, is at the heart of the turbulence problem. The limit of resolution occurs typically through the temporal and spatial resolution of observing systems, or the spatial resolution of numerical models but there is an intellectual challenge there too; I speculate that if we really understand a process, then we ought to be able to represent the transfer due to it parametrically, i.e. deduce what effect it will have in some given situation without having to follow the details of the behaviour.

Advective transfer can be represented as the product of a speed of advection and the lifetime of the property being transferred. The lifetime will depend on the scale of the motion. If the lifetime is very short the property will diffuse away before the advection is complete. If the lifetime is very long, particles will not give up their property. The lifetime depends on physical processes. Most significantly, on several occasions we have noted that momentum can be redistributed by pressure forces as well as by advection and diffusion. It may therefore be more readily transferred than those properties that can only be advected and diffused. Vorticity may be more conserved than momentum because it cannot be changed by pressure forces. Momentum and vorticity are both very 'active' variables in the sense of being closely associated with the advecting motion itself. For example positive and negative vortices may tend to go off in pairs, thus making the total transfer of vorticity small compared to some other property, like sensible energy.

Water vapour and sensible energy are more passive, though sensible energy may be a source of buoyancy, so not quite as passive as water vapour. Water vapour is almost passive until it condenses when latent energy is converted into sensible, as cloud particles grow. Cloud particles may subsequently evaporate, taking back the sensible energy from the air, or they may coagulate into raindrops and fall out. In these respects water vapour has a distinct lifetime compared with sensible energy. Neither are quite passive because they represent a source of thermal energy hence buoyancy, that affects the motion.

By way of illustration, the mean meridional circulation of Section 15.3 carries latent energy differently to the sum of latent and sensible, in the sense that latent energy is released as the air ascends but is thereby converted into sensible plus potential energy. One consequence is that the transport of latent energy by the mean meridional circulation is large compared with the transport of latent + sensible + potential energy. Large-scale motion is constrained by thermal–wind balance which constrains the upper-level wind to be nearly along the isotherms, so advection of temperature by large-scale motion is highly constrained.

Thus we need to consider the physical nature of the property being transferred as well as that of the motion system.

14.2 Mixing

Early theories of turbulent transfer laid much emphasis on the notion of mixing, to such an extent that some writers confuse the concept of transport with that of mixing. The notion of transport by mixing results in a formulation for transport analogous to the treatment of diffusion in the kinetic theory of gases in which theory a molecule is supposed to be discrete with a free path, a speed, and a lifetime terminated by a collision with another molecule. In this collision all properties of the two molecules are supposed to be substantially shared. The mean free path is a geometric property, defined as the distance a molecule is likely to travel before it meets another. The molecular speed dictates how quickly that path will be completed. Their product gives the molecular diffusivity. Such motion is sometimes referred to as random which it is not. We need to be careful because if it were truly random, there would be no correlation between the molecular velocity and the amount of property possessed by the molecule, and so no transfer. Correction can be made to distinguish between the transfer of molecular kinetic energy (thermal conductivity of heat) and momentum (viscosity) to take account of the different weighting of the averaging process for the speed, but this correction is comparatively small. Similarly for the persistence of some properties after the collision, and the diffusion of molecules of different size, culminating in the very slow diffusion of giant molecules as in Brownian motion.

14.3 Unresolved transport is not always mixing

It is possible to confuse the notion of advection by unresolved fluid motion, and that of mixing. If we monitor the flux of a particular property across a test surface, we can perform the correlation; fluctuating property density mutiplied by fluctuating normal velocity, and calculate the 'flux density' which defines the transfer. Some of this transfer may be due to bulk motion of the air and some may be due to mixing.

For example, a chimney plume will be carried by the wind as a meandering region in which the density of the smoke is large relative to the surrounding air. Mixing may take place at the plume boundaries, thereby increasing the volume of smoky air as it passes down the plume. The mass of the smoke does not of course increase. Transport of smoke, like whether it will obscure the sun at a given point, is likely to be dominated by the wandering of the plume boundary. But chemical reactions between different components of the smoke or between smoke chemicals and the outside air, is dependent through their number densities on the molecular separation of these constituents, and therefore on their molecular diffusion. Such chemical reactions depend little on where the plume happens to be.

Mixing is a specific mechanism, in which a property from one side of a test surface is shared irreversibly with that from the other, usually at a rate comparable with that of other properties.

Sometimes the nature of transport is obscured by the ploy of defining an eddy diffusivity. Thus we observe the flux density, divide by the gradient of the property being transferred, and so define a diffusivity. This quantity seems to define all we want to know about the transfer. But how do we predict the diffusivity defined this way, or rationalise its behaviour. In the kinetic theory of gases, the diffusivity can be expressed in terms of the bulk properties of mean free path, and speed of sound, but this association with bulk properties is less true of transfer in more complex systems and of other physical processes. For example, such mixing coefficients measured in systems of typical stable stratification, are several orders of magnitude greater in the horizontal than in the vertical. Also, as in Sections 11.9 and 15.5, for motion on the scale of weather systems a mixing coefficient for the horizontal transfer of momentum might well be of the opposite sign to that for the transfer of thermal energy and neither conveniently expressed in terms of a local gradient.

Thermal energy, momentum, water vapour, and carbon dioxide are atmospheric variables of some interest. Let us consider how they are transferred.

14.4 Transfer of energy

Thermodynamic equations for a parcel of air define those processes that govern the conservation of energy. We need to take account of sensible energy, latent energy, work done by pressure, gravitational potential energy, absorption and emission of electromagnetic radiation.

Molecular diffusivity is important in a region near the ground, where net radiation at material surfaces is converted into thermal energy by heating or cooling the surface and into latent energy by evaporation and sometimes condensation. This energy is passed by molecular diffusion to a thin layer of air close to the material surface, which soon reaches equilibrium. Further transfer into the air depends on this boundary layer being peeled away from the surface and replaced by another. The rate at which this transfer takes place hardly depends on the temperature and humidity difference between the surface and the air nor on the molecular diffusivities. It is better to think of the transfers as determining the differences. As a simple illustration of this principle; the wall flexes as I lean on it, but only enough to provide the force necessary to prevent me penetrating it. It is usually unnecessary for me to calculate this displacement, unless, that is, the wall is very soft and yielding. Perhaps as when I sit on a soft chair holding a cup of tea? Molecular processes may play a role in determining the ratio between latent and sensible heat transfer, at least rather close to the surface. The behaviour of the wet/dry bulb hygrometer in a relatively unventilated enclosure like a Stevenson screen illustrates this.

Thereafter in our parcel's flight, molecular conductivity is probably insignificant, except as a detail of the rain formation process, and at heights of order 120 km where molecular diffusivities become comparable with eddy diffusivities because of the increase of molecular mean free path. In the formation of rain, molecular diffusivity is combined with the rest of cloud microphysics to determine the size of cloud particles and rain drops.

Ordinary air is sufficiently unclean that it never becomes appreciably supersaturated so far as the thermodynamics is concerned. We can suppose that the latent energy is released at thermodynamic saturation and immediately transferred to the air. For this purpose we need not refer to microphysical details like the temperature difference between the particle and the air, nor to the diffusion of water vapour to, and sensible energy away from, the growing particle, nor to the fact that the supersaturation is of order 10^{-4} (another pure number that mathematicians would describe as being of order unity) which dictates how long it takes to grow a cloud droplet. However, if we want to know how big the droplets might be then we do need to represent these processes, and some others too. Cooling of air by evaporation of rain as in cumulonimbus downdraughts again demands knowledge of some molecular processes, though these can also be parameterised, albeit with less certainty than for the condensation process.

A typical global timescale for water vapour in the atmosphere is obtained by dividing the total amount of water vapour in a column of atmosphere which is some $50 \, \mathrm{kg \, m^{-2}}$ by the flux of water vapour at the ground, which is in turn equivalent to precipitation of 1 m per year, giving 20 days as the average lifetime of water vapour in the air, a life that is terminated by precipitation. Doing a similar calculation for air near a convective cloud gives a lifetime several orders of magnitude smaller. These timescales tell us something about the equivalent of a mean free path for water vapour, and the nature of the relevent physical processes.

How about thermal energy? Electromagnetic radiation, mostly in terrestrial wavelengths, results in the free atmosphere cooling by about 1 K per day. The thermal reservoir is not the 250 K of absolute temperature, because that is not attainable by meteorological processes, but the few tens of K of potential temperature change that is. This gives a thermal lifetime of a few tens of days.

From these values, and our knowledge of the flow structure, we begin to picture weather systems as transferring warm air quickly and adiabatically, precipitating excess water vapour, from one latitude and height to another. There it waits patiently, sinking gently, in the radiative deficit of the polar regions, becoming cold dry surface air. It is then rapidly transported, by a mid-latitude weather system, to low latitudes where it again waits patiently, this time being warmed and moistened by vertical convection. This may be a convoluted process. Perhaps the air first circulates round a well-mixed layer, taking turns to visit the ground and pick up moisture and sensible heat. Then it may be involved in a deep big cumulus event, where it shares its energy with the environmental air by forcing it down thereby warming it adiabatically, probably evaporating rain into it too.

On satellite pictures of the tropics we can sometimes see several old 'air masses' separated by feeble fronts which we might take as evidence of this process in action. Air in polar regions also seems to retain, for this sort of timescale, chemical evidence of its origins, like fine Saharan dust, and industrial carbon whose origin is defined by its chemical composition.

14.5 Transfer of momentum

Dynamical variables, like momentum or vorticity are difficult to treat because they are directly related to the motion carrying them about. If momentum were usefully conserved by parcels of air then the momentum equation would be nearly $Dv/Dt = 0$. When variations in one direction are small, we can expect the component of momentum in that direction to be more nearly conserved. This is certainly true of the zonal-mean flow where pressure cannot contribute net torque round the Earth's axis. It is also likely to be true of the 'logarithmic' layer observed near the ground, as suggested in Section 6.10 , where it is argued that the flow is in the form of rolls elongated in the down-shear direction, but in general, momentum is not conserved.

In several special cases some component of vorticity is rather well conserved, $D\zeta/Dt = 0$. Numerical models of the breakdown of a jet, intended to simulate nearly-geostrophic eddies in the Gulf Stream, generate eddies with positive and negative vorticity. These generally move around, not too far from the jet axis, slowly diffusing into each other, depending on how much sub-grid-scale diffusion there is in the model. Occasionally, however a pair will elope. This happens when they form a neat ($+/-$) vortex pair and induce corresponding velocities in each other. Perhaps it is just because there is no net transfer of vorticity that this is allowed: remember that the circulation round an isolated vortex is associated with logarithmically infinite kinetic energy. However, patches of fluid with compensating vorticity may carry other properties, like heat, with them, so because they are ineffectual at transferring vorticity they may be very efficient at transferring heat.

Perhaps no particle about to go on a long journey should be supposed typical of those surrounding it, which are not. The vortex experiment of Section 6.2 is salutory for it is a peculiar transport theory that will deduce that an initially uniform flat vortex sheet will become transformed into a row of fairly concentrated vortices.

14.6 Chemicals

Observations of the partial pressure of carbon dioxide in air and sea show that CO_2 is transferred from the sea to the air in low latitudes, from the air to the sea in high. There is also an annual variation of mixing ratio of CO_2 in the air, as it is extracted by plants in the spring, and returned in the autumn. The chemical lifetime of CO_2 in the air is a few months for this process, and it is advected by, and stored in, the atmosphere by motion of this timescale. In the ocean, the partial pressure changes mainly because the temperature of the parcel of water changes, but this takes a comparatively long time because of the slowness of ocean currents and the large thermal inertia of the mixed layer. Thus the oceanic timescale is a few years.

Chemical reaction demands molecular collision. An important reaction for the formation of ozone contains a three-body reaction between molecular and atomic oxygen, and an arbitrary air molecule. The rate of this reaction therefore depends on the cube of the number density. When this reaction takes place in the upper atmosphere, in the ozone layer, molecular diffusion changes the number densities only rather slowly, at least because the gradients of number density are very small. However, we can rather easily move a parcel vertically over distances comparable with the density scale height but conserving the mixing ratio, by gravity and quasi-geostrophic waves. This increases the reaction rate by perhaps a factor of three compared with that for the average molecular densities.

The same three-body reaction takes place in photochemical smogs where it is associated with oxides of nitrogen emerging from the exhausts of cars. In that situation the ozone is in the air outside the exhaust plume, and the nitrogen is inside the plume. In this case the reaction depends almost completely on molecular diffusion bringing together the molecules. In practice this is aided by the turbulent intermingling of the exhaust plume and the environment. On several other occasions we have noticed advective intermingling of fluids with dissimilar properties. Now consider how molecular diffusion acts in such deforming flows.

14.7 Diffusion and shear

Suppose that ordinary Fickian diffusion with constant diffusivity K, takes place in flow with a unidirectional shear $u = ay$, $v = 0$. A passive tracer of density ρ satisfies

$$\left(\frac{\partial}{\partial t} + ay\frac{\partial}{\partial x}\right)\rho = K\left(\frac{\partial^2}{\partial x^2} + \frac{\partial^2}{\partial y^2}\right)\rho \tag{14.1}$$

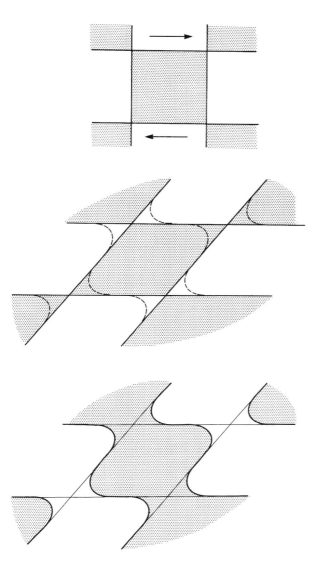

Figure 14.1 According to equation 14.2, diffusion in shear converts a checkerboard pattern first to stripes oriented in the opposite sense to the shear, then advects them to stripes directed down the shear. I find this sufficiently unexpected to demand a simple geometrical illustration. Having deformed the squares into lozenges, the corners get diffused away leaving upstream stripes.

Several aspects of diffusion become clear when one discusses sinusoidal initial distributions, like $\rho = A \exp \mathrm{i}(kx + \mu y)$, and we can generalise such solutions to include the effect of shear. Thus we expect that the property might tend to be carried along by the local sheared motion, which suggests that we look for a generalised solution of the form

$$\rho = A(t) \exp \mathrm{i}(k(x - ayt) + \mu y)$$

where

$$\mathrm{d}A/\mathrm{d}t = -K(k^2 + (\mu - kat)^2) A$$

or

$$\log A = -K\left(k^2 t + (kat - \mu)^3/3ka\right) \tag{14.2}$$

From this solution we see that for large time, shear increases the effective coefficient of diffusion by a factor $a^2 t^2$ as it compresses the y-extent of the wave.

Consider the effect of such shear and diffusion on a checkerboard pattern $\cos kx \cos \mu y$. Putting $\cos \mu y$ as a sum of $e^{+i\mu y}$ and $e^{-i\mu y}$ we see that the term containing $+i\mu y$ decreases from the start, relative to that containing $-i\mu y$. In this time interval, the pattern has tended towards $\rho = \exp i(kc + \mu y)$, or upshear stripes. This seems so astonishing that a geometric validation is attempted in Figure 14.1. Subsequently, the shear asserts itself and leans the stripes down-shear, as we would expect.

14.8 Diffusion and deformation

Another simple deformation is represented by the non-divergent flow

$$u = ax \qquad v = -ay \tag{14.3}$$

Integration following parcels gives

$$x = x_0 \exp +at \qquad y = y_0 \exp -at \tag{14.4}$$

so we might look for a solution like that for the sheared case except that the diffused property is a function of the Lagrangian coordinate $x \exp -at$, $y \exp +at$ rather than $x - ayt$. Indeed, we find in place of equation 14.2

$$dA/dt = -K\left(k^2 e^{-2at} + \mu^2 e^{+2at}\right) A \tag{14.5}$$

which we can easily solve for ρ. Diffusion in the y-direction is enhanced when $a > 0$, with a mutiplier $\exp 2at$, even more violent than the $(at)^2$ found for the shear-enhanced diffusion.

In each case, if molecular diffusion is to be substantially aided by the motion, there must not have been a significant amount of molecular diffusion by time of order $1/a$. Thus Kk^2/a must not be large. The inverse of this number defines a diffusion Reynolds number and for most meteorological scales this number is indeed large showing that diffusion is substantially assisted by deformation.

This is less true for smaller scales. This is because the shear and deformation are also limited by the similar diffusive action of molecular viscosity, so they too are dependent on the space scale; an aspect of their interactive nature. For the mechanically driven turbulence of Section 6.3, equation 6.9 suggests the dependency of shear on scale as being like $\partial u/\partial z \simeq v_m k^2$. This gives the diffusion Reynolds number as K/v_m, the ratio of the molecular diffusivities for molecular identity and momentum. For a disperse gas this is close to unity. In contrast, the diffusivity of soluble gases in liquid water is about three orders of magnitude smaller than that of momentum. Thus, in air, dispersion brings air of different origins fairly close together, though the final mixing is dominated by molecular processes. In water, the diffusion Reynolds number is of order 10^{-3}, so the deformation processes still leaves a large space gap to be covered by molecular processes.

Another interesting special case is where there is no viscosity at all, typified by the Helmholtz solution of Section 6.2. In that limit the length of the interface between the two fluids seems to increase linearly with time, suggesting more affinity with diffusion driven by shear than by deformation, where the interface lengthens exponentially with time.

Chapter 15

General circulation

We might see the objective of understanding meteorology, as getting a feel for why the planet has its observed mean temperature, which sets the field for water vapour transformations, clouds, latent heat, and life. A second objective might well be the assesment of fluctuations; the global and seasonal variation of winds and temperatures and humidities, which is what we are about to engage apon. Further objectives might well begin with persistent anomalies from the climate thereby established.

15.1 Definition of general circulation

Some set of mean overall properties of the atmosphere that change slowly with time are called the general circulation. Perhaps the most obvious of these sets, though not necessarily the easiest to explain, is the zonal mean of wind, temperature, humidity, etc. as functions of height. We find that some average variables are related to other average variables rather simply. For example it is difficult to justify a large imbalance between the zonal mean of the zonal component of the wind and the pole–equator temperature gradient, as constrained by the thermal wind. Other quantities depend almost entirely on the transport of various properties by eddy motion which may occur on a variety of scales. Thus the non-zero value of the zonally averaged surface wind depends almost entirely on the transport of momentum across latitude belts by eddies on the scale of weather systems. Similarly the relation between the net radiative heating imbalance, and the pole–equator temperature contrast depends almost entirely on the transport of heat by the weather systems. Wind in the first few metres above the surface depends on the local surface roughness. Variation of potential temperature with height must surely be dependent on the ensemble of (vertical) convective transport of heat.

Thus our general circulation must also contain such relevant statistical quantities if it is to be physically consistent. By implication it should also contain plausible physical interpretation of the important eddy transports.

15.2 Zonal mean observed

The height–latitude section up to about 20 km shown in Figure 15.1 shows many interesting features. We already have some misgivings about the nature of the averaging process. For example, what is the ground? Much of the data has been extrapolated to mean sea level, using a variety of physical principles of dubious validity, at least one of which results in about half the great Siberian

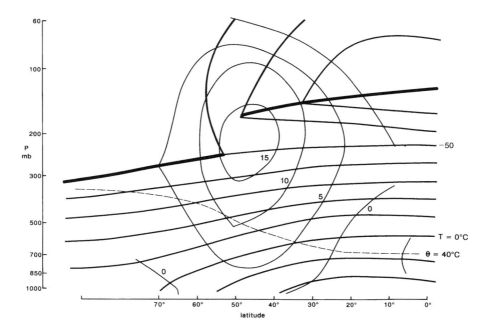

Figure 15.1 Zonal cross-section of the troposphere and lower stratosphere showing zonal component of wind, temperature, and potential temperature against height and latitude, in summer (above) and in winter (overleaf). We notice a troposphere of modest static stability, some 15 km deep in the tropics, 7 km deep in polar latitudes, with the temperature decreasing towards the poles almost everywhere, except perhaps just not for tropical latitudes near the ground. Wind strength is comparatively light near the ground increasing in

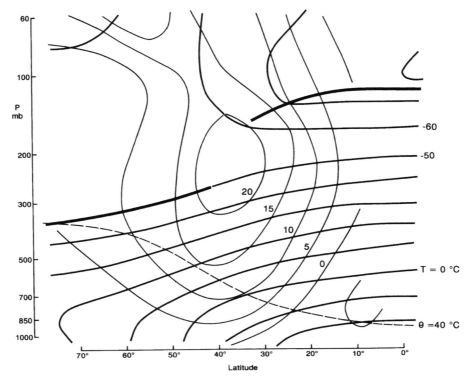

Caption for Figure 15.1 (*cont.*)
strength through the troposphere, and decreasing rather rapidly with height
into the stratosphere, associated with a general reversal of the pole equator
temperature gradient there. Surface winds are generally positive (west to east)
in middle latitudes, negative in the tropics, small or negative in polar regions.

winter anticyclone being below ground level. After this, the zonal average has
been taken.

High places are likely to be anomalous. Sinks of momentum are likely to be
concentrated in regions of high mountains, and therefore rough orography but
there are relatively few observing stations at high levels. There is not a great
area of land compared with sea, and much less high land but there are very few
stations in the middles of oceans, except of course where some, distinctly unrepresentative, steep-sided volcanic island has provided a site. Some meteorolo-
gists believe that this unrepresentivity is not numerically trivial, but most think
that the general picture is as we have described it.

15.3 Zonal mean dissected; thermodynamics

The zonal mean of the thermodynamic equation $D\phi/Dt = Q$ can be written in
flux form as

$$\frac{\partial}{\partial t}(\overline{\rho\phi}) + \frac{\partial}{\partial y}(\overline{\rho v\phi}) + \frac{\partial}{\partial z}(\overline{\rho w\phi}) = \overline{\rho Q} \tag{15.1}$$

where the overbar denotes the process of averaging with respect to x.

Our knowledge of baroclinic or slantwise convection leads us to expect, during the active phase of the convection, a horizontal flux of sensible heat $\overline{\rho C_p vT}$ of some $15\,\mathrm{kW\,m^{-2}}$, composed of 1 $\mathrm{kg\,m^{-3}}$ $10^3\,\mathrm{J\,kg^{-1}\,K^{-1}}$ $10\,\mathrm{m\,s^{-1}}$ $3\,\mathrm{K}$ $(1/2)$ where half is representative of the correlation between v and T, and vertical flux of some 15 $\mathrm{W\,m^{-2}}$ corresponding to (half) the isentropic slope multiplying the horizontal flux. There is a contribution in the same sense and of similar magnitude in the form of latent energy.

Our knowledge of vertical convection leads us to expect a vertical heat flux of some $5\,\mathrm{kW\,m^{-2}}$ composed of $10\,\mathrm{m\,s^{-1}}$ $1\,\mathrm{K}$ $10^3\,\mathrm{J\,m^{-3}\,K^{-1}}$ $(1/2)$ during the active phase of cumulonimbus, and a factor of 100 less for cumulus.

The flux carried by the mean circulation in the meridional plane is revealing. This circulation is most intense in the tropics where it is known as the Hadley cell. From the considerations of Chapter 8, we estimate the zonal mean *meridional* component of velocity in the tropics as some $1\,\mathrm{m\,s^{-1}}$ confined to the layer below about 1 km, and a return flow of $0.1\,\mathrm{m\,s^{-1}}$ through the remaining depth of the troposphere. This is not inconsistent with the observed values of Oort (1971), but there is argument about the representivity of data biased towards continental convection which gives regions of large meridional speed at high levels as well. Continuity of mass demands that the mass flux should very closely cancel when integrated with respect to pressure. A similar calculation to those above shows the horizontal transfer of sensible plus potential energy to be about $25\,\mathrm{kW\,m^{-2}}$ polewards composed of $1\,\mathrm{m\,s^{-1}}$ $25\,\mathrm{K}$ 10^3 $\mathrm{J\,m^{-3}\,K^{-1}}$ where the difference in potential temperature of $25\,\mathrm{K}$ is between the ground and halfway up the troposphere. We notice that flux of sensible energy $C_p vT$ is in the opposite sense; equatorwards, but note that it is not a properly useful thermodynamic quantity.

However there is also a flux of latent energy, and with the surface air some $10\,\mathrm{g\,kg^{-1}}$ moister than the upper air, this latent energy is almost the same size as the sensible energy but is being carried in the opposite sense. Thus we visualise the mean meridional circulation in the tropics as a device for converting latent into sensible-plus-potential energy, rather than achieving a net horizontal flux of their sum. Now the mean circulation is observed to give a net horizontal flux, but the value is rather uncertain and depends critically on how the sums were done; whether mixing ratio's have been averaged, or relative humidity, or wet bulb temperatures, for example.

Summarising, we expect significant contributions to the horizontal flux of latent plus sensible energy from the baroclinic scale, particularly in middle latitudes, dominance of cumulus in the vertical flux up to 1 km and comparable vertical flux from baroclinic waves and from cumulonimbus in the middle and upper troposphere. The ascending branch of the mean tropical circulation may well take place entirely in cumulonimbus towers. Indeed the mean circulation of the tropical air outside cumulonimbus might well be in the opposite direction.

Do we have to worry about this? A cumulonimbus extending all the way round a circle of latitude would conserve axial angular momentum, whereas a row of cumulonimbus spread around a line of latitude need not. Considerations broached in Chapter 7 suggest that we might not even be able to give the *sign* of the transfer of angular momentum in such a row of cumulonimbus. Many models like to have air ascend vigorously at a few gridpoints, and have to be dissuaded from doing this artificially. The discussion of Chapter 7 showed that there might be good reasons for this behaviour.

15.4 Zonal mean dissected; zonal component of momentum

The zonal mean of the zonal component of the momentum equation can be written

$$\frac{\partial}{\partial t}(\overline{\rho u}) + \frac{\partial}{\partial y}(\overline{\rho v u}) + \frac{\partial}{\partial z}(\overline{\rho w u}) - \overline{\rho f v} = \frac{\overline{\partial \tau_x}}{\partial z} \tag{15.2}$$

where we have chosen to separate out the transfer of zonal momentum due to small-scale motion into a stress τ, but without being too precise about what we intend by small scale nor what the implied averaging process is. Again the first term is small. Horizontal transfer of relative momentum in the second term is dominated by baroclinic motion. Vertical transfer of relative momentum in the third term is probably dominated by mechanical turbulence in the first 100 m, then by convective motion on the scale of cumulus up to heights of about 1 km and probably by cumulonimbus and gravity waves above, though mesoscale motion may not be negligible. The term fv representing the transfer of absolute momentum by the mean meridional circulation is large. In the lowest 1 km there is near balance between the vertical transfer and the mean circulation

$$\rho f v \simeq \tau_x/h \text{ or } v \simeq 0.1 \,\mathrm{N\,m^{-2}}/(1\,\mathrm{km}\ 1\,\mathrm{kg\,m^{-3}}\ 10^{-4}\mathrm{s^{-1}}) = 1\,\mathrm{m\,s^{-1}} \tag{15.3}$$

Above that region there is near balance between the momentum transfer by the mean circulation and the momentum transfer by the baroclinic motion. Momentum transfer by cumulonimbus might be significant but even if it is, it is more likely to carry momentum against the mean gradient (as in Chapter 7) rather than to mix it. Surface drag transmitted upwards by propagating gravity waves is also a significant source of momentum to the upper atmosphere.

The interest of this manipulation is emphasised when we integrate with respect to height from the ground, where $w = 0$ and τ is large, to the 'top' of the troposphere where w and τ are comparatively small. The mass flux ρv integrated with respect to height and latitude very very nearly vanishes because the mass of air in a column cannot be systematically accumulated. Thus, just as for the transfer of sensible + gravitational-potential + latent energy, the mean circulation serves to transfer momentum between the top and bottom of the troposphere, but cannot supply net momentum to the column. Here then is posed one of the puzzling features of the general circulation. Westerly momentum must be transferred out of the tropics, where it is being injected continuously through the action of surface friction which diminishes the surface winds which are negative relative to the surface (or easterly). This positive (or westerly) momentum must be transferred to middle latitudes where the surface winds are positive (or westerly), and provide a sink of westerly momentum. Thus the horizontal transfer of momentum is in the opposite sense to the gradient of mean momentum, i.e. it is counter-gradient transfer. This is now generally accepted to be associated with the transfer of potential vorticity down gradient, by the baroclinic eddies, or as a consequence of energy dispersion from middle latitudes, which amounts to much the same thing. It has been argued that the gradient of *angular* momentum is of the same sign throughout, therefore the transfer is not reversed in expected sign. This seems to me to negate all the advantages of writing down the momentum equations in terms of relative velocities in the first place. If we do use relative velocities, we discover this anomaly in momentum transfer, and that then demands explanation. We notice in passing, that had we written down the geostrophic momentum equation, which is commonly used (as in Chapter 9) to describe many aspects of large-scale motion, and then taken the zonal average, none of the products responsible for the eddy transfer of momentum would have appeared. As with the deception of the hydrostatic equation of Section 7.1 we need to recognise that what was the dominant variable in the W–E component of the local momentum equation $\partial p / \partial x$, vanishes when integrated round the latitude circle, so our criterion for the neglect of terms smaller than it is invalid for this particular purpose. A similar issue occurs in the 'Sverdrup' balance in the ocean circulation (Section 8.4.1).

15.5 Zonal mean dissected; meridional component of momentum

$$\frac{\partial}{\partial t}(\overline{\rho v}) + \frac{\partial}{\partial y}(\overline{\rho v v}) + \frac{\partial}{\partial z}(\overline{\rho w v}) + \overline{\rho f u} + \frac{\partial p}{\partial y} = \frac{\overline{\partial \tau_y}}{\partial z} \qquad (15.4)$$

As for the zonal component, the first term is utterly negligible for our present purpose. The momentum transfer represented by the second and third terms and by the stress, are probably somewhat smaller than the similar terms in the zonal component. However, the mean zonal wind u is at least a factor of 10 greater than the mean meridional component v so the dominant balance must be with the pressure gradient; the geostrophic approximation. More significantly, we can use the temperature gradient deduced from heat balance to determine the thermal wind shear derived from equation 15.4. This leaves the surface wind to be found from consideration of momentum balance.

We interpret the role of the mean transverse circulation as the mechanism for restoring thermal wind balance after it has been disturbed by sources and sinks of momentum and heat.

15.6 Horizontal transfer of momentum

The quasi-geostrophic momentum equation implies that we should not expect momentum to be transferred by large-scale eddies. Thus in

$$\rho D\mathbf{v}/Dt + \rho f\, \mathbf{k} \wedge \mathbf{v} + \nabla p = 0 \qquad (15.5)$$

the statement that a particle will take its momentum with it as it moves is that $D\mathbf{v}/Dt = 0$. But the quasi-geostrophic approximation implies that the balance is between the second and third terms, and that it can make no statement about $D\mathbf{v}/Dt$ except that it is small compared with say fv. Substituting realistic numbers one sees that this is a very crude statement for the leakage of momentum from a parcel of air, which supports our argument. The danger sign to watch for is to beware that we take one equation and make it into two by including the term we threw away as part of its structure. Only if the Coriolis and pressure gradient forces were small, as perhaps in the logarithmic boundary layer, should we expect particles to conserve their momentum as they moved from one place to another. For similar reasons, we would not expect angular momentum to be transferred either. When we look for dynamical quantities that *are* more nearly conserved we arrive first at the vertical component of absolute vorticity. Thus, as in Section 9.3, if the motion is very nearly horizontal, absolute vorticity $\zeta + \beta y$ will be transferred across lines of latitude by the motion.

Now examine the vorticity generated by the surface stress. Between 15 and 45° of latitude the surface wind has negative vorticity, so friction must be injecting positive vorticity here, and maybe in the polar regions, and negative at all other latitudes. The mean gradient of absolute vorticity is dominated by the β-effect and is everywhere larger towards the pole. Thus any mixing of mean vorticity will tend to bring (positive) vorticity equatorwards, away from the frictional source in latitudes 45–60° and into the sink in latitudes 0–15° which is good, but in the opposite sense for balance in the latitude belt 15–45°. In contrast, we have seen that quasi-geostophic eddies are even better at conserving and transporting quasi-geostrophic potential vorticity $Z = \zeta + \beta y + f\, \partial\phi/\partial z$. The additional term $f\, \partial\phi/\partial z$ is a consequence of the vertical stretching of air columns as they move. This is determined by the slantwise nature of the convection, and therefore with the baroclinic nature of the atmosphere. Thus though the transfer of potential-vorticity, and kinematic-vorticity are the same in the barotropic zones of high and low latitudes, the transfer may differ significantly in the baroclinic zones of middle latitudes.

The x-average flux of quasi-geostrophic potential vorticity is given by

$$\overline{vZ} = \frac{\partial}{\partial y}(\overline{uv}) + \frac{\partial}{\partial z}(\overline{v\phi}) \tag{15.6}$$

where non-divergence of the horizontal wind, and thermal wind balance for the zonal component have been invoked to remove terms that integrate out to zero round a latitude circle. This equation is encouraging because it enables the desired momentum flux $\overline{uv}$ to be found from the fluxes of plausibly conserved quantities. Unfortunately, the terms in heat and potential vorticity are about five times as large as the momentum flux, which implies finding the term we want as a small residual. However, we can demand that the angular momentum integrated over the globe should nearly balance. This is just enough to determine that the residual term does change sign, and, as we have seen this is likely to be in the baroclinic zone.

I thought that this scheme was very clever when I first saw it, but became disillusioned subsequently. What is most disturbing from a physical standpoint is that the transfers in equation 15.6 take place at different stages in the lifetime of the baroclinic wave. Thus the heat transfer certainly takes place in the amplifying phase, and becomes small and of uncertain sign at the mature stage, but the momentum transfer takes place in the mature phase, where the velocities are large or even later, in the dispersion phase. Thus we can imagine that the transfer coefficients giving the fluxes in terms of the mean gradients are significantly different for potential vorticity and heat making their small difference of uncertain accuracy.

15.7 A model of the general circulation; parameterisation

We might think of representing each of the important physical processes explicitly. Resolution of at least 100 m that one would need to represent cumulus convection explicitly is likely never to be available on a global scale. Gridlengths of 100 km have been used to resolve weather-scale systems. Resolution of 10 km is possible, though this suffers from the disadvantage of just-not-properly resolving mesoscale motion. Thus it seems that whatever resolution we choose, we need to represent an important group of smaller scales implicitly. This means that their important transport properties, which affect the large-scale motion, must be expressed in terms of the parameters of the large-scale motion itself. We could imagine that vertical transfer of heat by cumulus convection might be expressed in terms of the variation of potential temperature with height. This is not as good as might seem at first sight. On a day with convective activity we often find that the mean lapse rate is stable, i.e. with potential temperature increasing, though rather gently, with height. What happens is that on a day of convective activity, locally and instantaneously, the air near the ground becomes warmer than that in the deep boundary layer. This air then ascends, rather quickly transporting the warm air to the top, and allowing cool air to come to the ground. Since this process is rapid and localised compared with the radiative heating rates, the mean stratification will be stable for most of the time. We forsee that this sporadic character will be found in a wide range of circumstances. There will be cloud when the humidity on the large scale is only 80% or whatever. It will rain in a cloudy grid square even when the average cloud in that square is incapable of growing rain, sunshine will come to the ground through an overcast sky, shearing instability is hard to initiate but not so hard to maintain, and so on. This sort of statement, can form the basis for parameterisation. Thus we might suppose that convection will break out when the stratification on the large scale attains some given, slightly stable, value; the magnitude of the transport of heat to depend on the difference between the model stratification and the critical value. Another tactic is to suppose that the convection is so efficient as to override all other diabatic effects and give a characteristic stratification. This has the disadvantage that when convection just starts up the stratification will jump immediately to a different value. This gives a local shock, and creates large horizontal gradients between points that do and do not have convective activity. An important consideration is that we are really trying to represent the effect of all the physics that cannot be carried explicitly. Thus in the present example, we are trying to represent not only the transport of heat and moisture on all scales smaller than that of the gridbox but also the transfer of energy by electromagnetic radiation in terrestrial and solar wavebands. All these together make up the process of vertical redistribution of latent and sensible heat in the boundary layer and I suspect that we might be able to model their combined effect better than the individual contributions: by invoking a characteristic temperature structure in order to redistribute the heat anomaly with height. Sometimes new data shows that one of the component parts of a model can be improved, but when it is improved the model is worse than before. This could be because the deficiency in the process has been compensated for by adjusting another ill-known component to give a more acceptable result.

Finally we must decide on the range of scales, in time as well as space, that are to be represented parametrically. For the simplest model we can try to parameterise even the systems on the scale of the mid-latitude weather systems. These may be the easiest as well, because it is generally supposed that they are more-or-less continually active, so somewhat less sporadic and therefore more predictable, than they might be. But then suppose that some really important events like those responsible for generating, maintaining, or disrupting blocking anticyclones are triggered by unusual behaviour of the weather systems. If this is so then it is the anomaly in the normal behaviour that we need to predict. That the weather tomorrow will be the same as the weather today is a good guide to the weather to expect, but it is not a prediction in the sense that it is what we should expect with no knowledge at all.

15.8 Some simple models of the general circulation

We believe that it is warmer in the tropics than at the poles because more solar energy arrives there on unit surface area, but more radiative energy in the terrestrial wavelengths leaves the tropics at the ground because it is warmer there. Supposing that the Earth behaved like a perfect radiator, we might deduce that the ratio of tropical to polar re-radiation at a fixed height nearly balanced the solar input; the hypothesis that each latitude belt is close to radiative equilibrium. This model would give a plausible explanation of the surface temperature difference between equator and pole.

However this would be a poor model because the atmosphere is nearly opaque to radiation at terrestrial wavelengths, and this opacity is due almost entirely to water vapour. The critical mass of water vapour that makes the optical depth for this stream of radiation unity, is near the tropopause in the warm moist tropical latitudes, near the ground in the dry polar regions but nearly the same temperature at each. We trace this to the fact that the density of water vapour depends greatly on temperature and not much on the range of relative humidities, when each are characteristic of the observed atmosphere. This argument needs a little adjustment, because of the 30% or so of radiation in the water vapour window between the pure rotation and the vibration rotation absorption bands that escapes to space relatively easily in cloud free areas, so the terrestrial flux at the top of the atmosphere should vary a little with latitude, as is observed.

To a first approach to the truth the net input of energy to the combined atmosphere and ocean depends on the geometry of the solar beam, and on the albedo, but not much on the terrestrial radiation. The temperature difference between the tropics and poles is sufficient to generate the motion that can carry this excess thermal energy; the system is in nearly a steady state in this respect. As we have seen there are at least three mechanisms by which this might take place. The mean meridional circulation efficiently converts between latent and sensible and potential energy, but does not efficiently carry their sum, except possibly in very low latitudes, where there is not much energy to be carried anyway. Wave motions avoid the constraint of having to carry angular momentum, which is what inhibits the zonal mean circulations. Both stationary and transient waves may carry heat. Transient waves do it out of necessity, and observations suggest that they are the dominant mechanism. There are logistic difficulties because travelling waves that persistently develop in one area can leave behind a signal that looks like a stationary wave, so we think that if anything, the observed transfer of energy by stationary waves is an overestimate of the power of the physical mechanism.

If we knew the law of heat transfer, i.e. how the magnitude of the flux of heat is related to the temperature difference, we could find the temperature field that corresponds with the postulated heating. Section 9.5 suggests the law of transfer.

$$\overline{\rho\, c_p v\, T \delta\phi} \simeq \rho\, C_p\, T\, (g/4B)^{1/2} (\Delta y)^2 (\partial\phi/\partial y)^2 \tag{15.7}$$

There is some argument about a coefficient of proportionality to be inserted in this relation, which I have argued is not critical, and is nearly a universal constant. Using this relation together with the inferred heat flux from radiative imbalance, we get good agreement for the pole–equator temperature contrast and its seasonal variation in the troposphere.

This works quite nicely for the troposphere but the stratosphere is a problem. The temperature of the stratosphere is probably determined by a close approach to radiative equilibrium modulated by a mean meridional circulation. Radiative equilibrium is a consequence of intense radiative transfer in the 11 μm carbon dioxide absorption band, which is stronger than any other absorber at this level; water vapour having been left far below in the troposphere. The repercussions are not too clear. We see from the zonal cross-sections, that there is a pronounced reversal in the horizontal gradient of temperature just above the tropopause, which serves to confine the jet to near the tropopause. What seems odd to me, is that the jet should be so nearly just wiped out in the lower stratosphere. Indeed a number of very elaborate general circulation models run on very large computers have the same problem as I do, and the minimum value of wind above the tropopause turned out to be a rather poorly explained quantity. When this was recognised, various remedial measures were taken. Many of these were directed at the radiative balance in the lower stratosphere, some curious properties of aerosol were explored. Peter Sheppard used to say that when all conventional physics failed, some property of aerosol could always be invoked to get out of a difficulty, but whether these were cosmetic, or physical in origin is difficult to judge.

A hypothesis that I liked was that there are tiny winds quite often because this is the place one would expect the stationary waves to dump their zero-momentum as they propagated. We can read the same effect into the otherwise fortuitous low speed layer near heights of 100 km too. As the lower stratosphere became better observed, the areas of light wind have tended to become less, so the attractive, and the less attractive hypotheses to account for their existence have become less important.

We believe that thermal wind balance for the zonal component of wind is closely satisfied. Thus if we knew the surface wind we could integrate with respect to height to find the zonal wind at all heights. The surface wind is not large but, through frictional inflow, determines the mean meridional circulation, hence the main desert zones of the planet, so we would like to get it right. Momentum transfer may be more subtle than heat transfer, for it is not fundamental to the mechanism of the baroclinic waves. However, we can argue that these waves tend to transport potential vorticity between latitudes, and if we do we can account roughly for the pattern and intensity of the surface winds. Alternatively we can use the idea that momentum transfer will be towards the baroclinic zone; Davies' hypothesis, or towards the zone of generation of baroclinic waves; Eady's dispersion hypothesis. This would immediately give westerly surface winds in the baroclinic zones, and conservation of total torque ensure easterlies in low, and perhaps in high, latitudes. It is odd that, while this step is intellectually rather difficult and subtle, large numerical models seem to get it more nearly correct than simpler things like the growth of baroclinic waves, and the magnitude of the consequent heat transfer.

15.9 Stationary waves

The next degree of elaboration of the general circulation is the slowly changing, rather long waves. These are largely barotropic in character, in that wind shear and baroclinity are probably not essential for their maintenance. Rather it is topographic variation like unequal thermal response of land and sea, or orography that is the driving mechanism. Responses, like those explored in Chapter 11, behave much as expected. Surface cooling producing low-level anticyclonic circulation, upper level cyclonic, most intense on the downwind edge of the cooling anomaly.

Orographic ridges produce anticyclonic circulation through the depth of the atmosphere, most intense near the crest. A novelty is that propagation of wave energy into the upper atmosphere may be a substantial energy sink to the very long waves, as discussed in Section 11.6. Such theories work best when the systems are linear. Effects can be added together to give results. When magnitudes are typical of atmospheric circulation this additivity becomes less plausible.

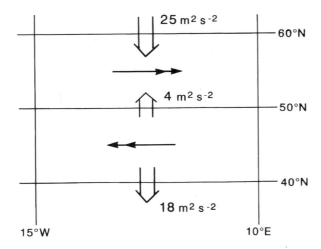

Figure 15.2 Box surrounding the British Isles showing the N–S transient eddy flux of zonal momentum $\overline{uv}$ in the upper troposphere. The jet was split to lie along 60° and 40°N. The eddy flux of momentum (open arrows) shows anticyclonic acceleration (double arrows) at upper levels. Momentum was transferred to the ground through a mean circulation with sinking at middle levels, bringing dry air, and momentum, to low levels, where it was balanced by a frictionally balanced anticyclone.

Another example where we need a large interactive model is when a blocking anticyclone develops. These typically have anticyclonic vorticity throughout the troposphere, and well into the stratosphere and are associated with rainfall anomalies. At least in retrospect they can be seen to have lasted for some many months, breaking down and reappearing spasmodically. The drought of 1976, which covered much of Western Europe with nearly half the normal rainfall, and was seen to have started in 1975, was a good example. Blocking refers to the apparent effect that the blocking anticyclone has in deflecting weather systems to the north and to the south of the block. One theory supposes that this is an essential feature for the maintenance of the block. Figure 15.2 shows a schematic picture. The mid-latitude baroclinic zone is split, and deflected to the north and the south by the anticyclone. Momentum transported by the weather systems developing in these zones is such as to inject anticyclonic vorticity into the anticyclone, particularly at upper levels. Anticyclonic vorticity is destroyed at the ground by surface friction and forces the descent of dry upper-tropospheric air to replaces the surface vorticity.

15.10 More general transfer

It is more complicated than that. Momentum equations in flux form equations 15.2 and 15.4, look nicer in terms of the dynamic pressure $p_* = p + \frac{1}{2}\rho v^2$ and are easier to manipulate:

$$\frac{\partial u}{\partial t} + \frac{\partial}{\partial x}\left(\frac{u^2 - v^2}{2}\right) + \frac{\partial}{\partial y}(uv) + \frac{1}{\rho}\frac{\partial p_*}{\partial x} - fv = \frac{\partial \tau_x}{\partial z} \qquad (15.8)$$

$$\frac{\partial v}{\partial t} - \frac{\partial}{\partial y}\left(\frac{u^2 - v^2}{2}\right) + \frac{\partial}{\partial x}(uv) + \frac{1}{\rho}\frac{\partial p_*}{\partial x} + fu = \frac{\partial \tau_y}{\partial z} \qquad (15.9)$$

Now we can eliminate the dynamic pressure to give a vorticity equation, in terms of the stresses $(u^2 - v^2)/2$ and uv. I do not know what the first of these is, and it worries me. I can perfectly see that u^2 is a physically acceptable x-stress, and algebraically we are left with the velocity correlation at $45°$, but that is not quite enough. We can plausibly take a temporal average of this set because the large-scale motion is relatively steady, and be left with temporally transient eddies generating the temporally steady large-scale motion. This expression contains terms representing the flux of vorticity, so why not write it down that way to start with? Indeed why not write it as the flux of potential vorticity that we think might be even more reliable? When we do we run into the problem that the potential vorticity is dominated by the comparatively large values of $f\,\partial\phi/\partial z$ in the stratosphere. But because the potential temperature of the stratospheric air *does* increase rapidly with height then surely it is going to be difficult to get that air to enter the troposphere and stay there! There is a growing body of evidence that what happens near the tropopause is a good indicator of development. But is this because it is the motion there which is important, or because we can easily see the effect there? Suppose we ran the model with a rigid lid near where the tropopause is? Well we cannot do that because it becomes impossible to analyse the initial data.

Another source of small potential vorticity, and so potential anticyclonic kinematic vorticity, is in the tropics, so why not bring tropical air into mid-latitudes and leave it in the anticyclone? An extension of the Azore's high the old timers used to call blocking anticyclones. Well perhaps the air would not stay there, like my stratospheric air might not.

Such climatalogical patterns are important, but to compare with observation it is necessary to use non-linear theory. Thus real orography takes up a con-siderable fraction of the troposphere, has steep slopes, and abrupt edges, and notices the variation of wind with height, hence knows about baroclinity. The details of climatalogical interest are of very small scale demanding exactitude; climatic shifts of interest to the Sahel drought are of order 100s of kilometres for example. All these conspire to force the use of general circulation models with their full physics of the large-scale motion and attendant and uncertain para-meterisations of the small-scale motion. This is new meteorology and is not the topic of this book. Given the new techniques that have been developed one becomes more subservient to the computer model, sometimes dependent for analysis of things one cannot observe but can deduce only from a detailed model, 'Ertel Potential Vorticity' being a good example. I get the feeling that we are studying the working of a computer model rather than the 'real atmosphere'. If we can run large general circulation models to diagnose hypotheses like these maybe the progress will continue. But what happens if we become so separated from observation and hypothesis that we become machine minders separate from the tangible atmosphere; the wind blowing in our faces, the long slog uphill against a head wind on our bicycle? What I have tried to present here is the power of tangibility; keeping in contact with everyday life. Maybe it is an unnecessary diversion. Quantum mechanics seems to have taken off just when, and because, it left off the idea of tangibility. Maybe meteorology will do the same?

> A rolling road,
> a reeling road,
> and such as we did tread,
> the night we went to Birmingham,
> by way of Beachy Head.

Appendix

Summary of data used in construction of Figures 1.1 and 1.2 some of which was only in report form

Figure 1.1
Curve 1: composite of analyses by Van Miegham, Benton and Kahn, Eliassen and Machenhauer, Saltzman and Fleicher, Winn and Nielsen, all for winter conditions.
Curve 2: Van der Hoven, 1957, *J. Meteorol.*, p. 160.
Curve 3: Garratt, 1969, PhD thesis, Imperial College.
Curve 4: Pinus 1967. *Tellus*, p. 206.
Curve 5: Panofsky and Deland, 1959, *Adv. Geophys.*, 6, p.49.
Curve 6: Buch and Panofsky, 1968, *Quart. J. R. Meteorol. Soc.*; T. Yanai, 1970, *Meteorol. Soc. Japan*; A. Sheppard, 1956, *Phil. Trans.*
Figure 1.2
Curve 1: Saltzman and Fleicher, 1960, *J. Geophys. Res.*
Curve 2: Buch and Panofsky, 1968.
Curve 3: Pinus 1967.
Curve 4: Kaimal and Haugen, 1967, *Q. J. R. Meteorol. Soc.*
Curve 5: Panofsky and Deland, 1959; A. Sheppard, 1956.

Index